ASTRONOMICAL APPLICATIONS OF
VEDIC MATHEMATICS

India's Scientific Heritage

General Editor: Dr L M Singhvi

7

Editorial Panel

ASTRONOMICAL APPLICATIONS OF
VEDIC MATHEMATICS

KENNETH WILLIAMS

Foreword by
L.M. SINGHVI
Formerly High Commissioner for India in the UK

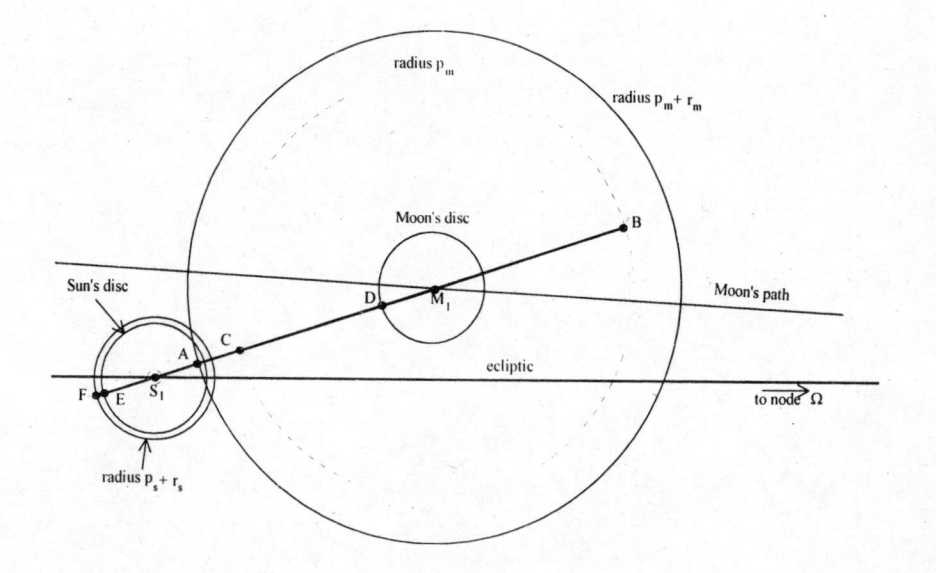

MOTILAL BANARSIDASS PUBLISHERS
PRIVATE LIMITED • DELHI

Reprint : Delhi, 2006
First Indian Edition : Delhi, 2003
(First Published by Inspiration Books)

ISBN: 81-208-1959-4 (Cloth)
ISBN: 81-208-1983-7 (Paper)

MOTILAL BANARSIDASS

41 U.A. Bungalow Road, Jawahar Nagar, Delhi 110 007
8 Mahalaxmi Chamber, 22 Bhulabhai Desai Road, Mumbai 400 026
236, 9th Main III Block, Jayanagar, Bangalore 560 011
203 Royapettah High Road, Mylapore, Chennai 600 004
Sanas Plaza, 1302 Baji Rao Road, Pune 411 002
8 Camac Street, Kolkata 700 017
Ashok Rajpath, Patna 800 004
Chowk, Varanasi 221 001

Printed in India
BY JAINENDRA PRAKASH JAIN AT SHRI JAINENDRA PRESS,
A-45 NARAINA, PHASE-I, NEW DELHI 110 028
AND PUBLISHED BY NARENDRA PRAKASH JAIN FOR
MOTILAL BANARSIDASS PUBLISHERS PRIVATE LIMITED,
BUNGALOW ROAD, DELHI 110 007

FOREWORD

Through trackless centuries of Indian history, Mathematics has always occupied the pride of place in India's scientific heritage. The poet aptly proclaims the primacy of the Science of Mathematics in a vivid metaphor:

यथा शिखा मयूराणां, नागानां मणयो यथा।
तद्वद् वेदांगशास्त्राणाम् गणितं मूर्ध्नि स्थितम्॥
—वेदांग ज्योतिष*

"Like the crest of the peacock, like the gem on the head of a snake, so is mathematics at the head of all knowledge."

Mathematics is universally regarded as the science of all sciences and "the priestess of definiteness and clarity". J.F. Herbert acknowledges that "everything that the greatest minds of all times have accomplished towards the comprehension of forms by means of concepts is gathered into one great science, Mathema-tics". In India's intellectual history and no less in the intellectual history of other civilisations, Mathematics stands forth as that which unites and mediates between Man and Nature, inner and outer world, thought and perception.

Indian Mathematics belongs not only to an hoary antiquity but is a living discipline with a potential for manifold modern applications. It takes its inspiration from the pioneering, though unfinished work of the late Bharati Krishna Tirthaji, a former Sankaracharya of Puri of revered memory who reconstructed a unique system on the basis of ancient Indian tradition of mathematics. British teachers have prepared textbooks of Vedic Mathematics for British Schools. Vedic mathematics is thus a bridge across centuries, civilisations, linguistic barriers and national frontiers.

Vedic mathematics is not only a sophisticated pedagogic and research tool but also an introduction to an ancient civilisation. It takes us back to many millennia of India's mathematical heritage. Rooted in the ancient Vedic sources which heralded the dawn of human history and illumined by their erudite exegesis, India's intellectual, scientific and aesthetic vitality blossomed and triumphed not only in philosophy, physics, astronomy, ecology and performing arts but also in geometry, algebra and arithmetic. Indian mathematicians gave the world the numerals now in universal use. The crowning glory of Indian mathematics was the invention of zero and the introduction of decimal notation without which mathematics as a scientific discipline could not have made much headway. It is noteworthy that the ancient Greeks and Romans did not have

---

* Lagadha, Verse 35

the decimal notation and, therefore, did not make much progress in the numerical sciences. The Arabs first learnt the decimal notation from Indians and introduced it into Europe. The renowned Arabic scholar, Alberuni or Abu Raihan, who was born in 973 A.D. and travelled to India, testified that the Indian attainments in mathematics were unrivalled and unsurpassed. In keeping with that ingrained tradition of mathematics in India, S. Ramanujan, "the man who knew infinity", the genius who was one of the greatest mathematicians of our time and the mystic for whom "a mathematical equation had a meaning because it expressed a thought of God", blazed new mathematical trails in Cambridge University in the second decade of the twentieth century even though he did not himself possess a university degree.

I do not wish to claim for Vedic Mathematics as we know it today the status of a discipline which has perfect answers to every problem. I do however question those who mindlessly deride the very idea and nomenclature of Vedic mathematics and regard it as an anathema. They are obviously affiliated to ideological prejudice and their ignorance is matched only by their arrogance. Their mindset were bequeathed to them by Macaulay who knew next to nothing of India's scientific and cultural heritage. They suffer from an incurable lack of self-esteem coupled with an irrational and obscurantist unwillingness to celebrate the glory of Indian achievements in the disciplines of mathematics, astronomy, architecture, town planning, physics, philosophy, metaphysics, metallurgy, botany and medicine. They are as conceited and dogmatic in rejecting Vedic Mathematics as those who naively attribute every single invention and discovery in human history to our ancestors of antiquity. Let us reinstate reasons as well as intuition and let us give a fair chance to the valuable insights of the past. Let us use that precious knowledge as a building block. To the detractors of Vedic Mathematics I would like to make a plea for sanity, objectivity and balance. They do not have to abuse or disown the past in order to praise the present.

Dr. L.M. Singhvi
Formerly High Commissioner for India in the UK

PREFACE

Since the publication of the book "Vedic Mathematics" by Sri Bharati Krishna Tirthaji in 1960 many new applications of the Vedic system have been found. Vedic Mathematics contains many examples of striking methods of calculation and there is a remarkable coherence to the system which makes it very attractive. Also the Vedic system itself suggests a kind of approach that involves going directly to the answer.

Vedic mathematics is based on sixteen Sutras and some sub-Sutras which provide links through mathematics: the word 'Sutra' means 'thread'. These Sutras are given in word form, for example *Vertically and Crosswise* and *By Addition* and *By Subtraction*, and where they arise in this text they are indicated by italics. The Sutras, and sub-Sutras, can all be related to natural mental functions. A full list of the Sutras and sub-Sutras will be found on Page 141.

Having had a keen interest in Astronomy for many years I had the opportunity, when studying for a degree in this subject, to look into Astronomical Applications of Vedic Mathematics for a final year project (in 1981). The result was what is now the contents of Chapters 2 and 3 of this book. The left to right method of calculation, which has so many useful applications (see "Vertically and Crosswise", Reference 3), was initially developed by the author in order to solve Kepler's Equation. Being encouraged to publish this work and take it further by studying for a higher degree, more such applications were found.

The methods given in this book are not intended to be a complete or thorough treatment of the topics they deal with. All of the ideas can probably be developed further or applied in other areas, and all can doubtless be improved upon. The mathematician will also observe a certain lack of rigor as an attempt has been made to make the material intelligible to as wide a readership as possible. To this end a Glossary and an Index have been added. An attempt has also been made to make the book as self-contained as possible so that the first chapter introduces some of the Vedic methods of calculation which are used in the book and the fourth chapter introduces the arithmetic for Pythagorean triples which is used in the subsequent chapters.

K. R. W.
January 2000

CONTENTS

Chapter 1

INTRODUCTION TO VEDIC MATHEMATICS

Vedic Mathematics was rediscovered from the 'Vedas' by Sri Bharati Krishna Tirthaji (1884-1960) between 1911 and 1918. The word 'Veda' literally means 'knowledge' and also refers to ancient Sanskrit writings contained in millions of manuscripts scattered over India. These texts contain details on a huge variety of subjects, including Mathematics. The system of Vedic Mathematics reconstructed by Sri Bharati Krishna presents a much more unified, direct and coherent account of mathematics which is much easier to learn and to do.

For the purposes of this book we will use the left to right method of calculation. Normal methods only carry out division from left to right; addition, subtraction and multiplication are done from the right. But there are many disadvantages to working only from the right and in the Vedic system we can work either way. Once left to right calculation is understood it is easy to combine operations and, for example, square two numbers, add them and take the square root (to any number of figures) in a single operation. In the calculation of trigonometric and other functions the Vedic system allows the valuations to be made from the left, first the most significant digit, then the next most significant and so on until the desired accuracy is obtained.

This chapter introduces the left to right method and its notation and shows the calculation of products, squares and division. The notation for cross-products, duplexes and the vinculum is used in the last part of section 2.3 and in Chapter 3.

1.1 PRODUCTS AND CROSS-PRODUCTS

(1) For the product of two 1-figure numbers we simply multiply the figures; this is a vertical product.

$$\begin{array}{r} 2 \\ 3 \times \\ \hline 6 \end{array}$$

(2) For the product of two 2-figure numbers we
   (a) take the vertical product on the right: 2.1 = **2**;
   (b) take the cross-product: 4.1 + 2.3 = **10**, put down 0 and carry 1;
   (c) take the vertical product on the left: 4.3 = **12**,

then 12 plus the carried 1 = 13.

$$\begin{array}{r} 4\ 2 \\ 3\ \ 1 \times \\ \hline 1\ 3{,}0\ 2 \end{array}$$

Thus **42 × 31 = 1302**.

We can see that this simple and symmetrical procedure works because the product of the units digits in the sum will give the units digit of the answer, the product of tens times units and units times tens in the sum (plus any carried figures) will give the tens digits of the answer and the product of the tens digits in the sum (plus any carried figures) will give the hundreds digits of the answer.

In carrying out this sum we obtained 3 **vertical products or cross-products**.

We can write:

$$CP\,_3^4\binom{2}{1} = 2, \quad CP\binom{4\;\;2}{3\;\;1} = 10, \quad CP\binom{4}{3}_1^2 = 12,$$

Please note that in this chapter we sometimes use a dot in place of a multiplication sign, e.g. $2.7 = 2\times7$.

(3) Similarly for the product $\begin{array}{c} 3\;\;2 \\ \underline{4\;\;7} \times \\ \end{array}$

the 3 cross-products are:

(a) $CP\,_4^3\binom{2}{7} = 2.7 =$ 1 4   i.e.   : |

(b) $CP\binom{3\;\;2}{4\;\;7} = 3.7 + 2.4 =$ 2 9   ✕

(c) $CP\binom{3}{4}_7^2 = 3.4 =$ $\underline{1\;\;2}$ | :

Therefore, $\underline{32 \times 47}$ =   1 5 0 4

(4) Now consider the product 302 × 514.
    Here we have 5 cross-products:

(a) $CP\,_{5\;1}^{3\;0}\binom{2}{4} = 2.4 =$ 8   : : |

(b) $CP\,_5^3\binom{0\;\;2}{1\;\;4} = 0.4 + 2.1 =$ 2   : ✕

(c) $CP\binom{3\;\;0\;\;2}{5\;\;1\;\;4} = 3.4 + 0.1 + 2.5 =$ 2 2   ✳

(d) $CP\binom{3\;\;0}{5\;\;1}_4^2 = 3.1 + 0.5 =$ 3   ✕ :

(e) $CP\binom{3}{5}_{1\;4}^{0\;2} = 3.5 =$ $\underline{1\;\;\;\;5}$ | : :

Therefore, $\underline{302 \times 514}$ =   1 5 5 2 2 8

(5) $3251 \times 7604 = \underline{24720604}$

$$
\begin{array}{cccc}
3 & 2 & 5 & 1 \\
7 & 6 & 0 & 4 \\
\hline
2\ 4\ _3 7\ _5 2\ _5 0\ _1 6\ _2 0\ 4
\end{array}
$$

or

$$
\begin{array}{cccc}
3 & 2 & 5 & 1 \\
7 & 6 & 0 & 4 \\
\hline
2\ 4\ _3 7\ _5 2\ _5 0\ _1 6\ _2 0\ 4
\end{array}
$$

We can write the answer straight down.

An alternative spacing is shown on the right. This has the advantage of enhancing the symmetry because each answer digit is put exactly under the centre of its product-pattern. Thus, for $5.4 + 1.0$ the centre is in the middle of these four digits and the answer goes below this point. As the sum proceeds these centre points move across the sum.

The cross-products, which can be evaluated mentally, are:

(a) $CP\ \begin{matrix} 3\,2\,5\, \\ 7\,6\,0\, \end{matrix}\!\left(\begin{matrix}1\\4\end{matrix}\right) = 1.4 =$ 4

(b) $CP\ \begin{matrix} 3\,2\, \\ 7\,6\, \end{matrix}\!\left(\begin{matrix}5&1\\0&4\end{matrix}\right) = 5.4 + 1.0 =$ 2 0

(c) $CP\ \begin{matrix} 3\, \\ 7\, \end{matrix}\!\left(\begin{matrix}2&5&1\\6&0&4\end{matrix}\right) = 2.4 + 5.0 + 1.6 =$ 1 4

(d) $CP\ \left(\begin{matrix}3&2&5&1\\7&6&0&4\end{matrix}\right) = 3.4 + 2.0 + 5.6 + 1.7 =$ 4 9

(e) $CP\ \left(\begin{matrix}3&2&5\\7&6&0\end{matrix}\right)\!\begin{matrix}1\\4\end{matrix} = 3.0 + 2.6 + 5.7 =$ 4 7

(f) $CP\ \left(\begin{matrix}3&2\\7&6\end{matrix}\right)\!\begin{matrix}5&1\\0&4\end{matrix} = 3.6 + 2.7 =$ 3 2

(g) $CP\ \left(\begin{matrix}3\\7\end{matrix}\right)\!\begin{matrix}2\,5\,1\\6\,0\,4\end{matrix} = 3.7 =$ $\underline{2\ 1\ }$

$$\underline{2\ 4\ 7\ 2\ 0\ 6\ 0\ 4}$$

(6) $12131 \times 20412 = \underline{247617972}$

$$
\begin{array}{ccccc}
1 & 2 & 1 & 3 & 1 \\
2 & 0 & 4 & 1 & 2 \\
\hline
2\ 4\ 7\ _6 1\ _1 7\ 9\ 7\ 2
\end{array}
$$

or

$$
\begin{array}{ccccc}
1 & 2 & 1 & 3 & 1 \\
2 & 0 & 4 & 1 & 2 \\
\hline
2\ 4\ 7\ _6 1\ _1 7\ 9\ 7\ 2
\end{array}
$$

There are 9 cross-products.

(7) 6341 × 32 = <u>202912</u>

If, as here, the numbers being multiplied do not have the same number of figures we can insert zeros:

$$\begin{array}{r} 6\ \ 3\ \ 4\ \ 1 \\ 0\ \ 0\ \ 3\ \ 2 \\ \hline 2\ 0\ _2 2\ _9 9\ _1 1\ 2 \end{array}$$

This method of multiplication is very easy and with a little practice it can also be very quick.

1.2 THE VINCULUM

Arithmetic operations can be greatly simplified by the use of the vinculum. In fact it is never really necessary in the Vedic system to use digits over 5 (i.e. 6, 7, 8, 9). This also provides flexibility as we can choose different ways of representing a number.

(8) As 38 is close to 40 we may use $4\bar{2}$ (i.e. $40 - 2$) instead of 38.

Similarly:

(9) $199 = 20\bar{1}$ (10) $2882 = 3\bar{1}\bar{2}2$ (11) $318297 = 32\bar{2}30\bar{3}$

(12) $278987 = 3\overline{21013}$ (13) $2\bar{6}341 = 14341$ (14) $34\overline{56} = 3344$

(15) $10\bar{2}2\bar{3} = 9817$

The Vedic formula ALL FROM 9 AND THE LAST FROM 10 helps us to get or remove these vinculum figures: in Example 12 the numbers under the vinculums are obtained by taking each of the large digits 78987 from 9 and the last from 10.

Similarly, in Example 14 we take 5 and 6 from 9 and 10 respectively to clear the vinculums.

In this way, not only are large figures avoided, but 0 and 1 appear twice as frequently as they otherwise would, and these are particularly easy digits to work with.

A third advantage is that numbers often partly or wholly cancel themselves, as will be seen below. We will be making extensive use of the vinculum.

(16) $39 \times 49 = \underline{1911}$ $39 \times 49 = 4\bar{1} \times 5\bar{1}$:

$$
\begin{array}{rr}
4 & \bar{1} \\
5 & \bar{1} \\
\hline
2 \quad 0 \quad \bar{9} \quad 1 \\
= \quad 1 \quad 9 \quad 1 \quad 1
\end{array}
$$

(17) $291 \times 388 = 3\bar{1}1 \times 4\bar{1}2 = \underline{112908}$:

$$
\begin{array}{rr}
3 & \bar{1} \quad 1 \\
4 & \bar{1} \quad 2 \\
\hline
1 \quad 2 \quad \bar{7} \quad \bar{1} \quad 1 \quad \bar{2} \\
= \quad 1 \quad 1 \quad 2 \quad 9 \quad 0 \quad 8
\end{array}
$$

(18) The vinculum is also very useful for division. For example, $41 \div 6$ is close to 7 so we may say that $41 \div 6 = 7$ remainder $\bar{1}$.

Similarly:

(19) $61 \div 9 = 7$ rem $\bar{2}$ (20) $13 \div 15 = 1$ rem $\bar{2}$ (21) $29 \div 6 = 5$ rem $\bar{1}$

1.3 LEFT TO RIGHT CALCULATIONS

It is often important, and especially useful in mental mathematics, to obtain the figures of an answer from left to right, i.e. to get the most significant figure first, then the next most significant figure and so on. Normally only divisions are done from the left.

Addition

(22)

$$
\begin{array}{r}
3 \quad 6 \quad 7 \\
9 \quad 8 \quad 5 \quad + \\
\hline
1_2 \qquad\qquad
\end{array}
$$

In this addition sum, if we add the left-hand column we get 3+9=12. Since there will be a carried figure from the next column that will affect this total, we put down only the 1 and carry the 2 forward,

as shown. The middle column adds up to 14. To this we add the carried 2, **as 20**, to get 34 and put down 3_4 as shown.

$$
\begin{array}{r}
3 \quad 6 \quad 7 \\
9 \quad 8 \quad 5 \quad + \\
\hline
1_2 3 \,_4 5 \quad 2
\end{array}
$$

Then adding the right-hand column we get 12, to which we add the carried 4, **as 40**, to get 52, which we put down.

(23)

$$
\begin{array}{r}
7 \quad 3 \quad 4 \quad 1 \\
2 \quad 3 \quad 8 \quad 7 \\
7 \quad 8 \quad 7 \quad 8 \quad + \\
\hline
1_6 7_4 \; 5_9 \; 10 \; 6 \\
1 \quad 7 \quad 6 \quad 0 \quad 6
\end{array}
$$

We could use the vinculum in this sum, and put $6\bar{1}$ instead of 59 in the third column:

$$
\begin{array}{r}
7 \quad 3 \quad 4 \quad 1 \\
2 \quad 3 \quad 8 \quad 7 \\
7 \quad 8 \quad 7 \quad 8 \quad + \\
\hline
1_6 7_4 6_1 0 \quad 6
\end{array}
$$

(24)   3   9   6
        8   8   7
        7   7   8   +
    $2_{\bar{2}}0_4 6 \ 1$

Subtraction

(25)   7 ^{1}3   2   In finding 732 − 184 we subtract in the left-hand column:
     1   8   4   −   7 − 1 = 6, but since a unit is needed for the next column
     5   (3 is less than 8) we put down only 5 and place the other
                          unit as shown.
     7 ^{1}3 ^{1}2   Then 13 − 8 = 5, but again a unit is needed in the next column
     1   8   4   −   (2 is less than 4), so we put down only 4.
     5   4   8   Finally 12 − 4 = 8.

(26)   5 ^{1}4 ^{1}7   2 ^{1}0 ^{1}3
     1   9   9   1   7   8   −
     3   4   8   0   2   5

Multiplication

(27)   3457 × 8 = <u>27656</u>

                   We begin on the left: 3.8 = 24, put as shown.
      3   4   5   7   Then 4.8 = 32: add the carried 4, **as 40**, 32 + 40 = 72, as shown.
              8   ×   Then 5.8 = 40: add the carried 2, as 20, 40 + 20 = 60, as shown.
    $2_4 7_2 6_0 5 \ 6$ Finally 7.8 = 56, 56 + 0 = 56, which we put down.

(28)   86379 × 6 = <u>518274</u>

                   Here we make use of the vinculum.
     8   6   3   7   9   8.6 = 48 = $5\bar{2}$, put $5_{\bar{2}}$.
                  6   ×   6.6 = 36, 36 + $\overline{20}$ = 16, put 1_6.
    $5_{\bar{2}} 1_6 8_{\bar{2}} 2_2 7 \ 4$ 3.6 = 18 = $2\bar{2}$, $2\bar{2}$ + 60 = $8\bar{2}$, put $8_{\bar{2}}$ etc.

(29)   1   6   8   8
              3   ×
    $0_3 5_{\bar{2}} 0_4 6 \ 4$

(30) $63 \times 74 = \underline{4662}$ Using the cross-product notation again:

$$\begin{array}{r} 6\ 3 \\ 7\ 4 \quad \times \\ \hline 4\,_2 6\,_5 6\ 2 \\ \hline \end{array}$$

$CP\begin{pmatrix} 6 \\ 7 \end{pmatrix} = 42$, put down 4 and carry 2.

$CP\begin{pmatrix} 6 & 3 \\ 7 & 4 \end{pmatrix} = 45$, and adding the carried 2, **as 20**, gives 65.

$CP\begin{pmatrix} 3 \\ 4 \end{pmatrix} = 12$, and adding the carried 5, **as 50**, gives 62.

(31) $\begin{array}{r} 4\ \ 3\ \ 7 \\ 5\ \ 2\ \ 6 \quad \times \\ \hline 2\,_0 2\,_3 9\,_5 8\,_2 6\ 2 \\ \hline \end{array}$

Here we find the cross-products:

$$\begin{pmatrix} 4 \\ 5 \end{pmatrix}, \begin{pmatrix} 4 & 3 \\ 5 & 2 \end{pmatrix}, \begin{pmatrix} 4 & 3 & 7 \\ 5 & 2 & 6 \end{pmatrix}, \begin{pmatrix} 3 & 7 \\ 2 & 6 \end{pmatrix}, \begin{pmatrix} 7 \\ 6 \end{pmatrix}.$$

(32) $\begin{array}{r} 2\ \ 4 \\ 3\ \ 2 \quad \times \\ \hline 0\,_6 7\,_6 6\ 8 \\ \hline \end{array}$

$CP\begin{pmatrix} 2 \\ 3 \end{pmatrix} = 6$ (a single figure), carry 6.

$CP\begin{pmatrix} 2 & 4 \\ 3 & 2 \end{pmatrix} = 16$, $16 + 60 = 76$.

$CP\begin{pmatrix} 4 \\ 2 \end{pmatrix} = 8$, $8 + 60 = 68$.

Using the Vinculum

---

(33) $\begin{array}{r} 8\ \ 2 \\ 6\ \ 8 \quad \times \\ \hline 5\,_{\bar{2}} 5\,_6 7\ 6 \\ \hline \end{array}$

Since $8.6 = 48$ we would prefer not to add 80 at the next step, so we can write 48 as $5\bar{2}$, put down 5 and carry $\bar{2}$

(34) $39 \times 69 = \underline{2691}$

$\begin{array}{r} 4\ \ \bar{1} \\ 7\ \ \bar{1} \quad \times \\ \hline 2\,_8 6\,_9 1 \\ \hline \end{array}$

Here we may change the figures in the sum to remove the large digits.

(35) Find 345243×761283 to 3 significant figures.

$\begin{array}{r} 3\ \ 4\ \ 5\ \ 2\ \ 4\ \ 3 \\ 7\ \ 6\ \ 1\ \ 2\ \ 8\ \ 3 \quad \times \\ \hline 2\,_1 6\,_{\bar{4}} 2\,_2 7\,_4 \ldots \\ \hline \end{array}$

$\therefore 345243 \times 761283 = 2.63 \times 10^{11}$ to 3 S.F.

1.4 MOVING MULTIPLIER

Multiplying a long number by a short number.

(36) 61261×43

$$
\begin{array}{c}
6\ 1\ 2\ 6\ 1 \\
\underline{4\ 3\qquad\qquad} \times \\
2_4 6_2
\end{array}
$$

We do not have to insert zeros when one number is shorter than the other: we can move the multiplier along its row, or imagine it to be moved, multiplying crosswise as we go.

First, in the position shown, the vertical product on the left is $CP\begin{pmatrix}6\\4\end{pmatrix} = 2_4$.

Then the cross-product $CP\begin{pmatrix}6&1\\4&3\end{pmatrix}$ gives 22 and 22 + carried 40 = 62 = 6_2.

$$
\begin{array}{c}
6\ 1\ 2\ 6\ 1 \\
\underline{\qquad 4\ 3\qquad} \times \\
2_4\ .\ 6_2 3_1
\end{array}
$$

Then we move the 43 one place to the right so that it is under 12 and $CP\begin{pmatrix}1&2\\4&3\end{pmatrix}$ = 11 + carried 20 = 31 = 3_1.

$$
\begin{array}{c}
6\ 1\ 2\ 6\ 1 \\
\underline{\qquad\qquad 4\ 3} \times \\
2_4 6_2 3_1 4_0 2_2 2\ 3
\end{array}
$$

We continue this process, moving the multiplier one place to the right each time and finding the new cross-product. In the final position we also take the vertical product on the right.

(37) Find 8273641×213 to 4 S.F.

$$
\begin{array}{c}
8\ 2\ 7\ 3\ 6\ 4\ 1 \\
\underline{2\ 1\ 3\qquad\qquad} \times \\
1_6 7_2 6_0 2\ _1 3_{\bar 4} \ldots
\end{array}
$$

Here we calculate from left to right; two shifts are needed.

$\therefore\ 1.762 \times 10^9$ to 4 S.F.

It is useful to know the number of zeros after the decimal point in a product when calculating from left to right.

(38) $0.008 \times 0.009 = 0.000072$

(39) 0.0812×0.032

$$
\begin{array}{c}
0.0\ 8\ 1\ 2 \\
\underline{0.0\ 3\ 2} \times \\
0.0\ 0\ 2\ 5\ 9\ 8\ 4
\end{array}
$$

We note from these two examples that the number of zeros after the decimal point in the answer is the same as the number of zeros after the decimal points in the numbers being multiplied.

(40) $0.004 \times 0.02 = 0.00008$

But here, since $4.2 = 8$, i.e. a single digit, there is one more zero in the answer.

1.5 SQUARING

We make extensive use of the "Duplex", D, for squares and square roots.
The term has the following meaning:

For a 1-digit number D is its square,

for a 2-digit number D is twice their product,

for a 3-digit number D is twice the product of the outer pair + the square of the middle digit,

for a 4-digit number D is twice the product of the outer pair + twice the product of the inner pair, and so on.

Thus $D(3) = 9$, $D(42) = 16$, $D(127) = 18$, $D(2134) = 22$, $D(21312) = 19$.

(41) For 43^2 we find the duplexes of the first figure, the pair of figures and the last figure:

$D(4) = 16$, $D(43) = 24$, $D(3) = 9$.

$\therefore 43^2 = 16/_24/9 = \underline{1849}$

(42) For 67^2 $D(6) = 36$, $D(67) = 84$, $D(7) = 49$:

$\therefore 67^2 = 36/_84/_49 = \underline{4489}$

(43) 341^2 $D(3) = 9$, $D(34) = 24$, $D(341) = 22$, $D(41) = 8$, $D(1) = 1$:

$\therefore 341^2 = 9/_24/_22/8/1 = \underline{116281}$

(44) 1435^2 We find the duplexes of 1, 14, 143, 1435, 435, 35 and 5.

$\therefore 1435^2 = 1/8/_22/_34/_49/_30/_25 = \underline{2059225}$

(45) 21034^2 We find duplexes for 2, 21, 210, 2103, 21034, 1034, 034, 34, 4.

The duplex of 21034 is $2.(2.4) + 2.(1.3) + 0^2$.

$\therefore 21034^2 = 4/4/1/_12/_22/8/9/_24/_16 = \underline{442429156}$

If we do the calculation from right to left we can take up the carry figures as we go.

1.6 DIVISION

The one line method of straight division works as follows:

(46) $3468 \div 72 = 48$ R12

$$
\begin{array}{r}
2\,)3\ 4\ 6'\ 8' \\
7\quad\ \ \ \ \ 6\,/\,2 \\
\hline
4'\ 8'\ 12
\end{array}
$$

We set the sum out as shown. Then $34 \div 7 = 4$ R6, as shown. Then 4 (in the answer) × 2 (raised ON THE FLAG) = 8, 66 – 8 = 58 and 58 ÷ 7 = 8 R2, placed as shown.
Next 8 (in the answer) × 2 (on the flag) = 16, 28 – 16 = 12 placed as shown.

It will be seen that these operations are identical to those in Example 47, but the procedure is simpler since at each stage we just multiply the last answer digit by the flag number, subtract the product from the next dividend, divide by the (single digit) divisor and put down the result.
The sum thus proceeds in distinct stages indicated by the diagonal lines above.

(47)

$$
\begin{array}{r}
1\,)2\ 1\ 3'\ 1'\ 4'\ 1' \\
8\quad\ \ \ \ 5\,/\,3\,/\,1\,/\,3 \\
\hline
2'\ 6'\ 3'\ 1|\,30
\end{array}
$$

As in the previous example, having one figure on the flag we mark off one figure on the right of the dividend to indicate the remainder column.

(48) Using the vinculum:

$$
\begin{array}{r}
3\,)3\ 7\ 3\ 7|\ 3 \\
6\quad\ \ \ \ 1\ \ \bar{5}\ |\ \bar{1} \\
\hline
6\ 0\ 7|\ 14
\end{array}
$$

$37 \div 6 = 6$ R1.
$6.3 = 18$, $13-18 = \bar{5}$, $\bar{5} \div 6 = 0$ R$\bar{5}$.
Our next dividend is $\bar{5}7$ or $\overline{43}$.
$0.3 = 0$, $\overline{43} - 0 = \overline{43}$, $\overline{43} \div 6 = \bar{7}$ R$\bar{1}$ as shown.
$\bar{7}.3 = \overline{21}$, $\bar{1}3 - \overline{21} = \bar{7} + 21 = 14$.

(49) Avoiding the vinculum:

$$
\begin{array}{r}
3\,)3\ 7\ 3\ 7|\ 3 \\
6\quad\ \ \ \ 7\ 4\ |2 \\
\hline
5\ 9\ 3|\ 14
\end{array}
$$

If we wish to avoid the vinculum then instead of
$37 \div 6 = 6$ R1, we may say
$37 \div 6 = 5$ R7 and continue as usual.

(50) Vinculum on the flag:

$$
\begin{array}{r}
\bar{1}\,)6\ 2\ 3\ 1|\ 2 \\
5\quad\ \ \ 1\ 3\ 0\ |3 \\
\hline
1\ 2\ 7\ 1|\ 33
\end{array}
$$

If when dividing by 49 we put 9 on the flag the subtractions are likely to be large.
If we use $5\bar{1}$ as a divisor instead of 49 we **add** the product of the last answer digit and the flag number at each step instead of subtracting.
Thus $6 \div 5 = 1$ R1, $\bar{1}.1 = \bar{1}$,
$12 - \bar{1} = 12 + 1 = 13$, $13 \div 5 = 2$ R3 etc.

(51) $54545 \div 29$

$$\overline{1})\underset{\quad 2\ \ 1\ \ 2\ \ \ 1}{5\ 4\ 5\ 4}\ \Big|\ 5$$

3

$$\underline{1\ 8\ 7\ \ 10\big|25}$$

$$=\ 1\ 8\ 8\ \ 0\ R25$$

Here we get 10 in the last place of the answer, so we carry 1 over to the left (but multiply the flag digit by **10** at the last step).

(52) Decimalising the remainder: $333 \div 73$

$$3)\underset{\quad\quad 5\ \ 6\ \ 3\ \ 5}{3\ 3\ 3\ .\ 0\ 0\ 0}$$

7

$$\underline{4\ .\ 5\ 6\ 1\ .\ .\ .}$$

The decimal point occupies the same position as the vertical line would have, and we just keep dividing in the usual way.

(53) 2-digit divisor: $776655 \div 203$

$$3)\underset{\quad\quad 17\ \ 7\ \ 12\ \ 19\ \ 0}{7\ 7\ 6\ 6\ 5\ 5\ .\ 0}$$

20

$$\underline{3\ 8\ 2\ 5\ .9\ .\ .\ .}$$

We can use a 2-digit divisor, 20 in this case, if we wish.

(54) Two figures on the flag: $2999222 \div 713$

$$13)\underset{\quad\quad 1\ \ 1\ \ 5\ \ \ 4\ \ 36}{2\ 9\ 9\ 9\ 2\big|\ 2\ 2}$$

7

$$\underline{4\ 2\ 0\ 6\big|\ 344}$$

$29 \div 7 = 4$ R1.
Then 4.1 (on the flag) = 4, $19 - 4 = 15$, $15 \div 7 = 2$ R1, as shown.
We then cross-multiply the two answer digits by the two flag digits: $1.2 + 3.4 = 14$, $19 - 14 = 5$, $5 \div 7 = 0$ R5, as shown.

Again multiply the last two answer digits by the flag digits: $1.0 + 3.2 = 6$, $52 - 6 = 46$, $46 \div 7 = 6$ R4, as shown.
Cross-multiply again: $1.6 + 3.0 = 6$, $42 - 6 = 36$, put as shown.
Then finally take the last (vertical) product of 6 (in the answer) $\times$ 3 (on the flag) = 18, $362 - 18 = 344$, the remainder.

(55) Three figures on the flag: $987987 \div 8123$

$$123)\underset{\quad\quad 1\ \ 1\ \ \ 5\ \ 51\ \ 510}{9\ 8\ 7\big|\ 9\ 8\ 7}$$

8

$$\underline{1\ 2\ 1\big|\ 5104}$$

$9 \div 8 = 1$ R1, put it down.
1.1 (on the flag) = 1, $18 - 1 = 17$, $17 \div 8 = 2$ R1, put it down.
$1.2 + 2.1 = 4$, $17 - 4 = 13$, $13 \div 8 = 1$ R5, put it down.

Next we cross-multiply the three answer figures with the three flag digits:
$1.1 + 2.2 + 3.1 = 8$, $59 - 8 = 51$, put it down.
$2.1 + 3.2 = 8$, $518 - 8 = 510$, put it down.
Finally $3.1 = 3$, $5107 - 3 = 5104$.

(56) Decimalised remainder: 54341 ÷ 7103

$$103)5\ 4\ 3\ 4\ 1.0\ 0\ 0\ 0\ 0$$
$$7 \quad \underline{5\ 4\ 3\ 5\ 4\ 0\ \overline{3}\ 0}$$
$$\underline{7.6\ 5\ 0\ 4\ 3\ 0\ \overline{6}..}$$

(57) Using the vinculum: 34567 ÷ 6918

$$\overline{1}2\overline{2})3\ 4\ 5\ 6\ 7.0\ 0\ 0$$
$$7 \quad \underline{6\ 6\ 4\ 4\ 3}$$
$$\underline{4.9\ 9\ 6\ 6\ 7...}$$

(58) Find 877778 ÷ 819976 to 5 decimal places

$$200\overline{2}\overline{4})8\ 7\ 7\ 7\ 7\ 8.0$$
$$8 \quad \underline{0\ 5\ 1\ 3\ \overline{1}\ 0}$$
$$\underline{1.0\ 7\ 0\ 5\ \overline{1}}$$

Chapter 2

PREDICTION OF ECLIPSES

An eclipse of the Sun occurs when the Moon obscures the disc of the Sun as seen from a position on the surface of the Earth. The Sun's disc may be partly obscured or totally. Figure 1, which is not to scale, shows the phenomenon. The area AB on the Earth's surface is a circular or oval-shaped patch (much smaller than indicated in Figure 1) in which the Sun is totally obscured by the Moon. This area of shadow (called the umbral shadow) moves across the surface of the Earth from west to east during an eclipse (see Figure 7). Observers in the areas indicated by BD and AC are in the penumbral shadow (partial shadow) and will see a partial eclipse. All other observers do not see an eclipse.

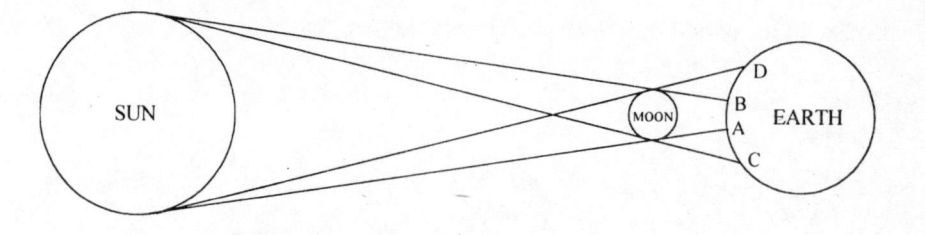

Figure 1: How a solar eclipse occurs
when the Moon partly or totally blocks the light from the Sun.

If the orbit of the Moon around the Earth was in the plane of the Earth's orbit around the Sun (called the ecliptic) the Moon would eclipse the Sun every month. The Moon's orbit is however inclined at about 5° to the ecliptic and so an eclipse only takes place when a New Moon occurs when the Moon is close to the ecliptic. We say the Moon must be near a node of the Moon's orbit, a node being where the Moon's orbit crosses the ecliptic (Figure 2). The angular size of both the Sun and the Moon is about half a degree.

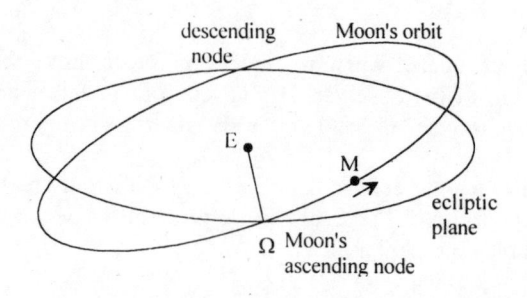

Figure 2: The Moon's inclined orbit and nodes.

2.1 PREDICTION OF THE TIMES OF CONTACT OF THE MOON'S PENUMBRAL AND UMBRAL SHADOWS WITH THE EARTH

Here we show how to predict the times of first and last contact of the Moon's umbral and penumbral shadows with the Earth during a solar eclipse. A quadratic equation is obtained for the penumbral and umbral phases, the solutions to which give the times of first and last contact. If there is no real solution to the first of these equations there is no eclipse of any kind at that conjunction of the Sun and Moon. No real solution to the second equation, after getting a real solution to the first, would indicate the eclipse is partial only. The total eclipse of 1981 July 31st will be used as an example throughout this chapter.

Figure 3 shows the relative positions and motions of the Sun and Moon as seen from the centre of the Earth on that day. No eclipse, partial or total, is seen from the Earth's centre. The three positions correspond to:

(1) start of the partial phase, with the Moon at M_1 and the Sun at S_1,

(2) conjunction (in longitude, i.e. M_2S_2 is perpendicular to $S_2\Omega$), and

(3) end of the partial eclipse phase, Moon at M_3 and Sun at S_3.

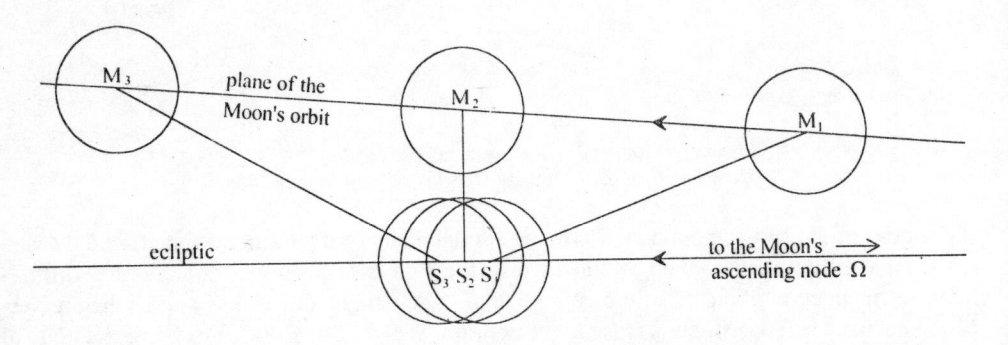

Figure 3: Relative positions of the Sun and Moon (seen against the fixed star background) at three times on 1981 July 31 as seen from the Earth's centre.

Parallax is the effect created when an object appears to move relative to an object further away due to motion of the observer. If you look out of the window and see a tree in the foreground and then move your head to one side the tree moves relative to its background.

The partial phase of a solar eclipse is when the Moon is only partly obscuring the Sun, also called the penumbral phase. The total eclipse is when the Moon completely covers the Sun and is also called the umbral phase.

Because of the proximity of the Moon any displacement of an observer at the centre of the Earth at right-angles to the line joining the Earth and Moon will produce an apparent displacement of the Moon relative to the stars: the parallax effect. Since the size of the Earth is limited there is a limit to the displacement that can be produced in this way. It follows that by suitable displacement of the observer on the earth's surface, the Moon's centre can be apparently displaced, relative to the stars, to any point inside a circle whose centre is that of the Moon as seen from the Earth's centre, and whose radius is equal to the equatorial horizontal parallax of the Moon, p_m (which is about 1°). This circle is equivalent to a mirrored image of the Earth and is therefore slightly oblate. There is also another circle with the same centre, and radius (p_m+r_m), where r_m is the Moon's angular radius, in which the disc of the Moon must always lie. Similarly two circles of radius p_s and (p_s+r_s) are associated with the Sun, which has a parallax, p_s, of about 9 arc seconds and radius r_s.

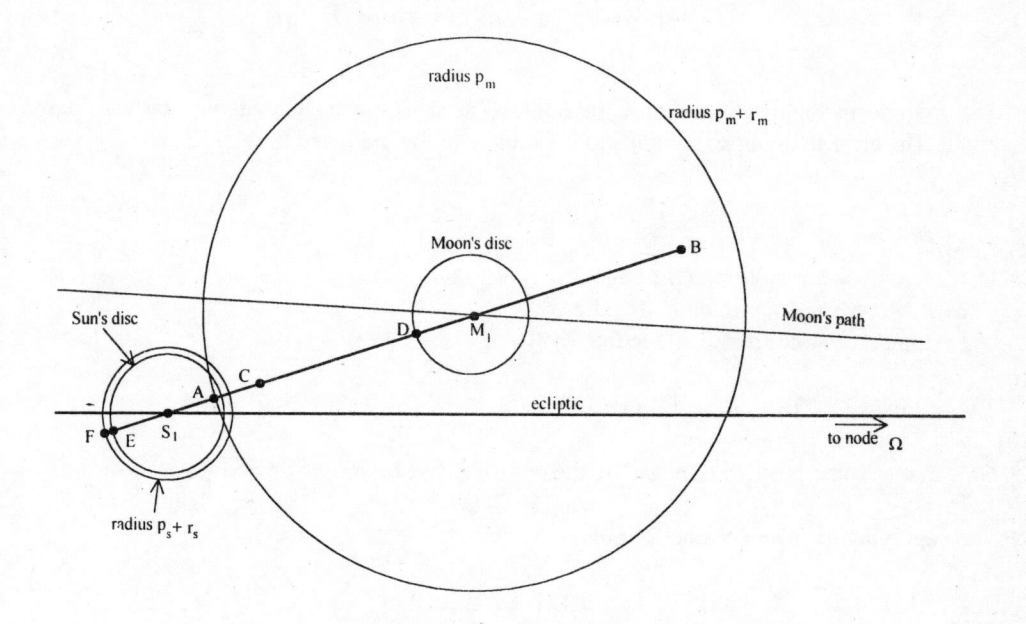

| Partial Phase |

Figure 4: The Sun and Moon as seen from the Earth's centre
at the start of the penumbral phase, showing the circles of parallax.

The diagram shows the position of the Sun and Moon against the stars as seen from the centre of the Earth at some time. The dotted circle of parallax also represents an image of the Earth were it at the Moon. For an observer on the Earth corresponding to a position B against the same background of stars, M_1 would appear to be at C, D at A, and E at F. The discs of the Sun and Moon would then be just touching and so B represents a position on the Earth of an observer who first experiences a partial eclipse. It will be seen that $S_1M_1 = (p_m+r_m-p_s+r_s) = s$.

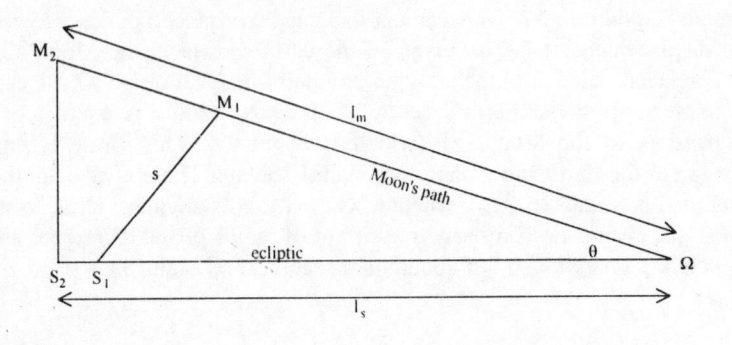

Figure 5: The positions of the centres of the Sun and Moon
at first contact of the penumbral shadow, S_1M_1, and at conjunction, S_2M_2,
relative to the Moon's ascending node, Ω.

The triangle in Figure 5 (in which the angle Ω has been exaggerated) will be considered plane. The error in doing so amounts to 0.1 minute in the predicted time.

Let $S_2\Omega = l_s$,
  $M_2\Omega = l_m$,
  n_s = Sun's mean motion in the ecliptic,
  n_m = Moon's mean motion in its orbit,
  t = time measured from conjunction (M_2S_2).

Therefore, $M_1\Omega = l_m + n_mt$ and $S_1\Omega = l_s + n_st$.

Note that t can be positive or negative, the negative case being shown above.

Then applying the plane cosine formula:

$$s^2 = (l_m + n_mt)^2 + (l_s + n_st)^2 - 2(l_m + n_mt)(l_s + n_st)\cos\theta$$

where θ is the inclination of the Moon's orbit to the ecliptic.

Therefore,

$$(n_m^2 + n_s^2 - 2n_mn_s\cos\theta)t^2 + (2l_mn_m + 2l_sn_s - 2n_ml_s\cos\theta - 2n_sl_m\cos\theta)t$$
$$= s^2 - l_m^2 - l_s^2 + 2l_sl_m\cos\theta.$$

With an error of thousandths of a minute of time the coefficient of t^2 can be written as $(n_m - n_s\cos\theta)^2$, so that

$$(n_m - n_s\cos\theta)^2t^2 + 2(l_mn_m + l_sn_s - \cos\theta(n_ml_s + n_sl_m))t = s^2 - l_m^2 - l_s^2 + 2l_sl_m\cos\theta. \quad \ldots (1)$$

The values of t obtained from equation (1) give the times of beginning and end of the partial phase of the eclipse, and if the roots are not real there is no eclipse. The equation will give the time of any desired separation of the Sun and Moon near a node (or, conversely, the separation at any time), and can therefore also be used to find the times of the total phase. For the eclipse of July 31st 1981:

$p_m = 58'.548$
$r_m = 15'.953$
$p_s = 0'.1443$ All four of these are from the Astronomical Almanac, 1981, Page A79.
$r_s = 15'.757$

$n_m = 0.5800'/min.$ This is the Moon's mean motion in its orbit and is obtained by finding the difference in longitude of the Moon at 3 a.m. and 4 a.m. and multiplying by sec θ.

There can be a difference of over $1°$ between the longitude of the mean node and that of the true node, so that the mean value cannot be used (unless an error of 1 minute in the result is permissible). The longitude of the true node is given in The American Ephemeris compiled by Neil

$n_s = 0.0398'/min$ Michelson, from which n_s (the Sun's mean motion in the ecliptic), and the remaining data are also drawn.

$T = 3^h53^m$ E.T.   This is the time of conjunction in longitude.

$l_s = 369'.24.$ This is the difference in longitude of the Sun and true node at conjunction (see Figure 5)

$l_m = 370'.80.$ This is found by getting the latitude of the Moon at conjunction $(34'.0)$ and applying Pythagoras Theorem to $\Delta S_2 M_2 \Omega$ in Figure 5 (i.e. $l_m^2 = l_s^2 + 34.0^2$).

$\theta = 5°.26102.$ This is the instantaneous inclination of the Moon's orbit to the ecliptic and is found from $\tan \theta = $ (Moon's latitude) $\div l_s$.

Putting this data into formula (1) gives the quadratic equation:

$0.2920t^2 + 3.6141t = 6964.5$ the solutions to which are $t = -160.75$ and $+148.37$.

Applying these to $T = 233$ minutes the times of beginning and end of the partial phases are found to be $1^h 12^m$ and $6^h 21^m$ E.T. (to the nearest minute).

From the Astronomical Almanac these times are found to be:

$1^h 12.2^m$ and $6^h 21.3^m$ E.T. (note that E.T. – U.T. = 0.87^m)

Total Phase

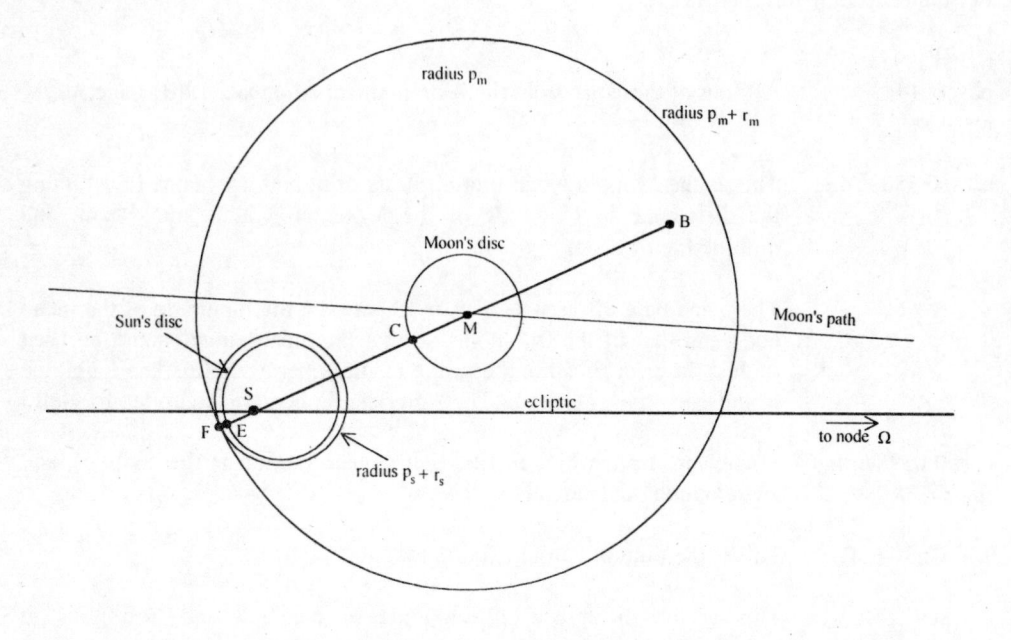

Figure 6: The relative positions of the Sun and Moon and the circles of parallax
at the start of the total eclipse phase.

The times of the total phase can be obtained in a similar way. The observer B (Figure 6) who first sees a total eclipse will see C and E at F, so that $SM = s = (p_m + r_m - p_s - r_s)$. Therefore, equation (1) gives: $0.2920t^2 + 3.6141t = 2277.9$, in which the left-hand side is the same as before.

This gives $t = -94.73$ and $+82.35$.

Applying these to T gives the times of the beginning and end of the total phase as:

$2^h 18^m$ and $5^h 15^m$ E.T.

These times are not given in the Almanac, but the times of contact of the central line (the line through the centres of the Sun and Moon) with the Earth's surface are given. This contact occurs at a later time than above, when SM is diminished by $(r_m - r_s)$.

In which case $s = 3410.99$.

Therefore, $0.2920t^2 + 3.6141t = 2254.91$, giving times of $2^h 19^m$ and $5^h 15^m$ E.T. (to the nearest minute).

The Almanac times are $2^h 18.7^m$ and $5^h 14.7^m$ E.T.

For a comparison of this method with the standard procedure the reader is referred to pages 233-235 of the Planetary Supplement to the Astronomical Ephemeris 1977 where the standard method is set out.

The following notes will be referred to in the next section.

Note 1. If in Figures 4 and 6 the dashed circle is thought of as an image of the Earth, then West is on the right and North is at the top.

Note 2. If S and M are the positions of the Sun and Moon during the total phase, and B is the corresponding point of contact of the central line, then $MB = SM - \frac{P_s}{P_m}SM$, i.e. $MB \approx SM$. In particular, MB (at conjunction in R.A.) $\approx$ SM (at conjunction in R.A.).

Note 3. If in Figure 6 S is considered stationary, as M moves along its path B traces out a line almost parallel to the Moon's path (actually, due to the Sun's motion the lines diverge at an angle of about $0°.2$).

2.2 THE APPROXIMATE POSITION OF THE ECLIPSE PATH

The 'Elements of the Eclipse' as given on page A79 of the Astronomical Almanac 1981 allow us to easily establish the approximate position of the path of totality on the Earth's surface. Figure 7 shows the Earth as it is seen from the Sun. From this direction the circle around which the Earth's north pole, N, appears to move in a year is seen as a straight line AC. Since the interval between the time of the summer solstice and the time of conjunction is known, the position of N can be found and is approximately as shown. The position of the equator and parallels of latitude are then also determined.

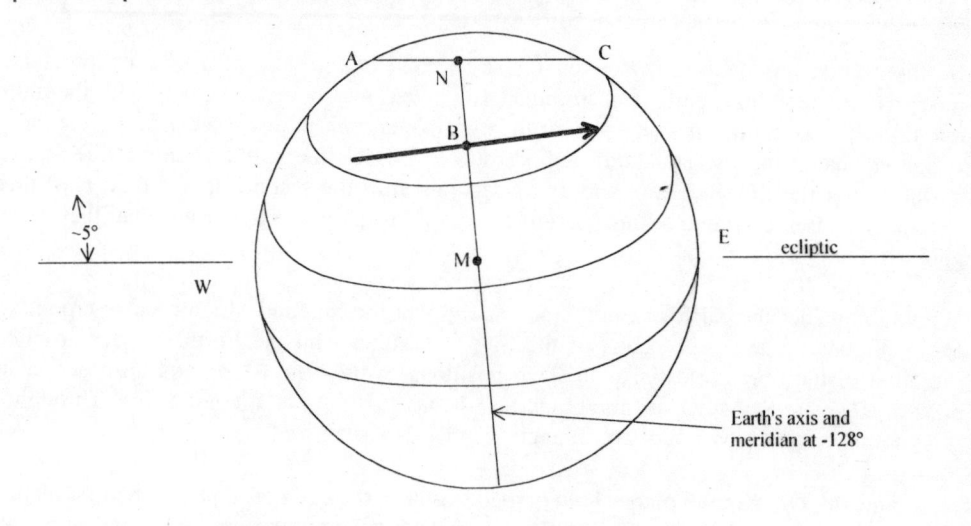

Figure 7: View of the Earth at the time of conjunction in R.A., as seen from the Sun, showing the approximate position of the eclipse path.

The longitude L of an observer on Earth who sees the eclipse at local apparent noon is given as −128° (this can easily be calculated from the time of conjunction: see section on Early Eclipse Prediction in this chapter), and the Earth's axis in Figure 7 would coincide with this line. Thus the hemisphere facing the Sun at conjunction is determined.

As the Moon's declination is greater than that of the Sun the eclipse path will fall in the northern hemisphere.

From Note 2 at the end of the previous section, MB (at conjunction in R.A.) ≈ SM (at conjunction in R.A.), and SM (at conjunction in R.A.) = difference of declinations of the Sun and Moon = 34'. And since, on the scale of Figure 6, the Earth's radius is p_m = 58' the position of B on NM is determined. In short MB (in units of the Earth's radius) ≈ difference in declination of Sun and Moon ÷ parallax of the Moon.

Finally, from Note 3 of the previous section, the orientation of the eclipse path θ is known to be about 5° to the ecliptic, and if it is also known that the Moon has passed its ascending node, the position of the eclipse path is thus approximately determined.

Since the given semidiameter of the Moon is greater than the Sun, the possibility of an annular eclipse can be ruled out.

Of the quantities used in this section to determine the eclipse path, p_m, θ, and the date of the solstice (or equinox) are virtually constants, so that the time and date of conjunction (from which L is derived) and the difference in declination of the Sun and Moon are the only quantities required to determine approximately the eclipse path.

2.3 · TIME OF TOTAL ECLIPSE FOR AN OBSERVER ON THE EARTH

In this section an equation is derived for calculation of the time of a total eclipse for an observer on the eclipse path. The formula, which can only be solved easily with the aid of one of the Vedic Sutras, has an error for the example chosen of just 5 seconds. By contrast it is shown that using the standard method, due to Bessel, the result obtained differs by 7 minutes after the first iteration, and by 12 seconds after the second iteration. At least three iterations of Bessel's method are therefore required to give a better result than that shown here.

We will consider the earth's centre is at rest and that the Sun and Moon revolve around the Earth. Figure 8 shows the orbits of the Sun and Moon, and the Earth, as seen from the direction of the north celestial pole. Two positions of the Sun, Moon and an observer are shown: these are S_t, M_t, O_t at time t, and S_0, M_0, O_0 at t = 0, i.e. at conjunction. Throughout this section conjunction refers to conjunction in Right Ascension.

Now consider two parallel planes both perpendicular to the equatorial plane: the first plane π_t always passes through the centres of the Sun and Moon (i.e. through the shadow axis), and the second plane Π_t always passes through the centre of the earth E (i.e. through the Earth's axis). At conjunction the two planes coincide and their separation s is then zero. As time passes the planes move apart and also rotate.

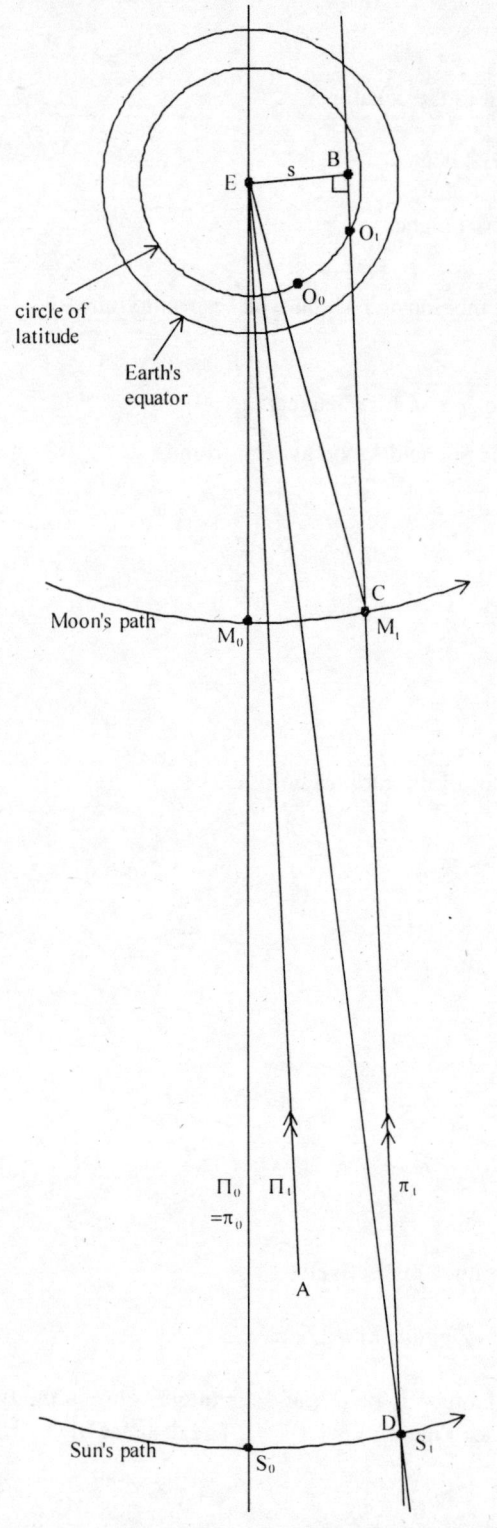

An observer O_t on the Earth is moving on his circle of latitude, and if he is to see an eclipse at all it will be when O_t lies in π_t, i.e. when the perpendicular distance s' of O_t from Π_t is equal to s. Expressions will be obtained for s and s' in terms of t, which will then be equated to give a formula from which t can be found. It will be seen that the problem of finding the time of the eclipse has been simplified by effectively eliminating motions perpendicular to the equatorial plane, and retaining only those in the equatorial plane.

Figure 8: The Sun, Moon and Earth at conjunction and at time t.

Separations of the Planes Π and π

Consider EM_t and ES_t as planes perpendicular to the equator.
B is the perpendicular from E to π_t.
C is the perpendicular from M_t to the equatorial plane.
D is the perpendicular from S_t to the equatorial plane.
A is on the intersection of Π_t with the equatorial plane.
δ_m is the declination of the Moon.

$\dot{\alpha}_s$ and $\dot{\alpha}_m$ are the rates of change of R.A. of the Sun and Moon with respect to time.

Then $E\hat{C}B = A\hat{E}D + D\hat{E}C$

Therefore, $E\hat{C}B \approx D\hat{E}C$ (see Note (1) at the end of this Chapter).

But $D\hat{E}C$ = angular separation in R.A. of the Sun and Moon as seen from E.

Therefore, $D\hat{E}C = (\dot{\alpha}_m - \dot{\alpha}_s)t$ at time t.

In ΔECB, $EC = \dfrac{s}{\sin E\hat{C}B}$.

In ΔECM_t, $EM_t = \dfrac{EC}{\cos \delta_m} = \dfrac{s}{\sin E\hat{C}B \cos \delta_m}$.

But $EM_t = \dfrac{1}{\sin p_m}$ (taking the equatorial radius of the earth as unity).

Therefore, $s = \dfrac{\sin E\hat{C}B \cos \delta_m}{\sin p_m}$.

Therefore, $s = \dfrac{\sin(\dot{\alpha}_m - \dot{\alpha}_s)t \cos \delta_m}{\sin p_m}$

Therefore, $s = \dfrac{(\dot{\alpha}_m - \dot{\alpha}_s)t \cos \delta_m}{p_m}$ (see Note (2)).

Distance s' of the Observer from Π_t

ϕ, ψ be the observer's latitude and longitude respectively,
ρ the distance of the observer from E,
L the longitude of conjunction, i.e. the longitude of Π_0 or π_0.

Then the radius of the observer's circle of latitude is $\rho\cos\phi$, and the angle between the two circles of longitude in Π_0 and through O_0 (see Figure 9) is $(L - \psi)$. The distance of O_0 from Π_0 is then given by $\rho\cos\phi\sin(L - \psi)$.

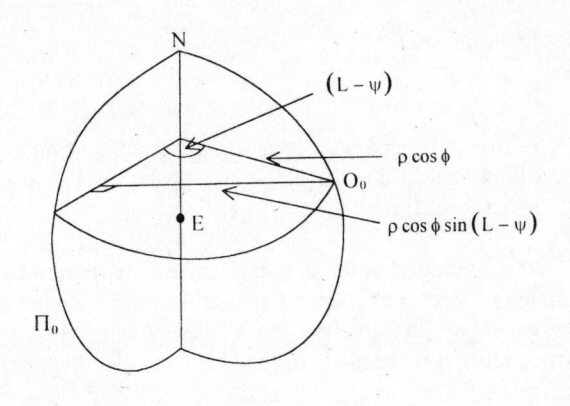

Figure 9: The two circles of longitude through Π_0 and O_0.

As time passes the observer moves on to O_t and the angle $(L - \psi)$ is increased by an amount equal to the rotation of the Earth lessened by the rotation of Π_t. The rotation of the Earth is 15 degrees per hour, or $\frac{1}{4}$ degrees in t minutes. The rotation of Π_t is due mostly (there is a small opposite rotation due to the Moon's motion perpendicular to the shadow axis) to the revolution of the Sun around the Earth, which amounts to $-\frac{\dot{a}_s t}{60}$ degrees per minute. This last term will be neglected, however, as it is small compared with the rotation of the earth. Also ρ will be taken as unity: these approximations are discussed in Notes (3), (4).

Then $s' = \cos\phi \sin\left(L - \psi + \frac{1}{4}\right)$.

For eclipse, $s = s'$ so that $\cos\phi \sin\left(L - \psi + \frac{1}{4}\right) = \dfrac{(\dot{\alpha}_m - \dot{\alpha}_s)t \cos\delta_m}{p_m}$,

and therefore $t = \dfrac{p_m \cos\phi}{(\dot{\alpha}_m - \dot{\alpha}_s)\cos\delta_m} - \sin\left(L - \psi + \frac{1}{4}\right)$.   ...(2)

Note that $\left(L - \psi + \frac{1}{4}\right) < 90°$

This equation will now be applied to finding the time of mid-eclipse for an observer at: $\phi = 35°.32$, $\psi = -177°.30$. According to the Astronomical Almanac 1981, Page A85, this observer would be in mid-eclipse at exactly $5^h 00^m$ U.T.

All the data required for equation (2) are immediately available on Page A79 of the Almanac:

$(\dot{\alpha}_m - \dot{\alpha}_s)$ = 145.026 – 9.741 seconds of arc per hour
= 0.5637 arc minutes per minute (see Note (5)).
$\delta_m = 18°.88$ (see Note (6)).
$p_m = 58.55$ arc minutes.

$L = -127°.69$ (though available, this quantity can be easily calculated—see later).

Therefore, $t = \dfrac{58.55 \cos 35.32}{0.5637 \cos 18.88} \sin\left(-127.69 + 177.30 + \frac{1}{4}\right)$.

That is, $t = 89.544 \sin\left(49.61 + \frac{1}{4}\right)$.

The solution of this transcendental equation is $t = 84.53$ minutes and since t is measured from conjunction which is at $3^h\ 35^m.56$ U.T. the eclipse time is found to be $5^h\ 00^m.09$ U.T. giving an error of 5 seconds of time.

The Vedic method of solving the above equation is explained at the end of this chapter (the method is different from that used in Chapter 3 to solve Kepler's Equation). Alternatively the equation can be solved iteratively using a calculator. Further remarks on the errors involved in the use of equation (2) are made in the Notes later in this chapter.

Comparison with Bessel's Method

The time of mid-eclipse will now be calculated for the same observer, but using Bessel's method. The calculation requires access to the table of Besselian Elements specially constructed for that eclipse, and availabie in the Astronomical Almanac, Page A84. A calculating device is also necessary, preferably one with several storage registers. The method proceeds as follows.

Make a first estimate T_0 of the time, say conjunction at $3^h\ 30^m$ U.T.

Therefore, $T_0 = 3^h\ 30^m$ U.T., $\phi = 35°.32$, $\psi = -177°.30$.

μ_0 (from the table) $= 230°.91940$ therefore $(\mu_0 - \psi) = 408°.2194 = 48°.2194$.

Also $\sin d = 0.314128$, and $\cos d = 0.949381$ and let $\rho = 1$.

Then using the usual formulae for Bessel's method:

$$\xi_0 = \rho \cos\phi \sin(\mu - \psi) = 0.608444687$$

$$\eta_0 = \rho \sin\phi \cos d - \cos\phi \sin d \cos(\mu - \psi) = 0.378104388$$

$$\zeta_0 = \rho \sin\phi \sin d + \cos\phi \cos d \cos(\mu - \psi) = 0.697733877$$

$$\xi_0' = \mu'(-\eta \sin d + \zeta \cos d) = 0.142342901$$
$$\eta_0' = \xi \sin d\mu' - \zeta d' = 0.050165925$$
$$\zeta_0' = -\xi \cos d\mu' + \eta d' = -0.15131233$$

where $\mu' = 0.261832$, $d' = -0.000175$ are also given on Page A84.

From the elements, $x_0 = -0.050817$, $y_0 = 0.595091$
and $x_0' = 0.547977$, $y_0' = -0.089034$ are obtained from the table by numerical differentiation.
Then $u_0 = x_0 - \xi = -0.05926168$, $u_0' = x_0' - \xi_0' = 0.405634099$,
   $v_0 = y_0 - \eta = 0.216986611$, $v_0' = y_0' - \eta_0' = -0.13919992$.

Therefore, $n_0^2 = (u_0')^2 + (v_0')^2 = 0.183915641$,
   $D_0 = u_0 u_0' + v_0 v_0' = -0.29762353$.

Therefore, $t_{mp} = \dfrac{-D_0}{n_0^2} = 1^h.61826 = 97^m.1$.

So the time of eclipse after the first iteration is $T_0 + t_{mp} = 5^h\ 7^m.1$ U.T.

This time is 7 minutes too late and to obtain a better result the above procedure has to be repeated. To save having to make several interpolations in the Besselian Elements $T_1 = 5^h\ 10^m$ will be taken.

Then $\mu_1 = 255°.92252$, $(\mu_1 - \psi) = 73°.22252$, $\rho = 0.998883$

   $\sin d = 0.313851$, and $\cos d = 0.949472$.

   $\xi_1 = 0.780331306$, $\xi_1' = 0.061598903$

   $\eta_1 = 0.474479947$, $\eta_1' = 0.064195497$

   $\zeta_1 = 0.404621917$, $\zeta_1' = -0.19407507$.

$\therefore$ $x_1 = 0.862342$, $y_1 = 0.446306$,
   $x_1' = 0.547794$, $y_1' = -0.089502$.

$\therefore$ $u_1 = 0.08201069$, $u_1' = 0.486195096$,
   $v_1 = -0.02817394$, $v_1' = -0.15369749$.

$\therefore$ $n_1^2 = 0.260008592$, $D_1 = 0.044203462$.

$\therefore$ $t_{mp} = -0^h.1700077 = -10^m.20046$.

Therefore, the time of eclipse after the second iteration is $T_1 + t_{mp} = 4^h\ 59^m.80$ U.T.

This time is 12 seconds too early. For greater accuracy a third iteration would be necessary, in which case the values extracted from the table of elements would have to be interpolated, making the calculation even longer.

As can be seen Bessel's method is long and tedious: it involves many formulae, and consequently many calculations, and requires access to the table of elements and a calculating device. It is however, capable of giving a very accurate result. By contrast, the formula put forward in this section (which is in error by only 5 sec) requires knowledge of just eight quantities including the observer's latitude and longitude, all of which are readily available, and which give an equation that can be solved quickly without the aid of any artificial calculating device.

Early Eclipse Prediction

It is interesting to consider whether the seven quantities used in equation (2):

$$t = \frac{p_m \cos\phi}{(\alpha_m - \alpha_s)\cos\delta_m} - \sin\left(L - \psi + \tfrac{1}{4}\right),$$ and T, the time of conjunction, might have been available

to astronomers of an earlier age. If they were, then it is possible that they could have predicted eclipses with some accuracy.

A brief outline follows of some methods by which these eight quantities may have been ascertained. Knowledge of the sphericity of the Earth is assumed.

An observer's latitude ϕ can be measured as it is equal to the angular distance along his meridian from the celestial equator to his zenith, or it is the complement of the angle between the north (or south) celestial pole and the zenith.

An observer's longitude ψ can be measured, relative to some convenient zero of longitude, if the observer and another on the zero longitude both noted which star was on their meridian at a certain instant. Then the angular distance between those two stars, measured along the celestial equator (i.e. their difference in right ascension), gives the difference in longitude of the observers. This difference in right ascension is more easily measured if the chosen stars are on or near the equator. Or, if the time is measured from this instant until the more westerly observer sees the other observer's star in the same position (relative to the horizon) as that observer saw it, then that interval will be the same fraction of a day as the difference in longitudes is of 360°. Hence, the longitude can be calculated by proportion. In both these methods the observers need to observe an event simultaneously. Assuming synchronised clocks are not available, they could wait for some celestial event to occur such as a very bright meteor or a lunar eclipse, and in the latter they would do better to use the moment of total disappearance or reappearance of the Moon as these moments are more accurately determined than the first or last contact of the Moon with the shadow.

Once the difference in longitude of the two observers is known, the distance can be measured around the observers circle of latitude between the two circles of longitude. From this the distance between the circles of longitude along the equator can be calculated (by dividing by the cosine of that observer's latitude), and hence the radius of the Earth can be determined. Now as the difference in longitude of the two observers is known, the difference in their local times is known, and they can therefore arrange to observe the Moon's position at the same instant, the stars being used as a clock. The difference in the two observed positions is due to the Moon's parallax. If the two observers were on the same circle of latitude the Moon's parallax could be measured when the Moon is on the meridian of a place midway between the two observers, and since the length of the straight line joining the observers can be calculated the parallax corresponding to the Earth's radius p_m can be found; the angle of parallax being so small that simple proportion can be used. A correction may be applied to allow for the fact that the centre of the line joining the observers is nearer to the Moon than is the Earth's centre.

That the angular size of the Moon varies, and therefore its distance and therefore its parallax, is apparent from the observation of annular as well as total eclipses. This variation is of the order of $\pm 6.5\%$ with a cycle of one (anomalistic) month. If variations of this order were detectable future values of the parallax could be predicted with greater accuracy.

Right Ascension is more easily and accurately measured than longitude. The Earth is a natural observatory, turning like a clock with almost perfect precision, once in a sidereal day. Instruments fixed to the earth could therefore measure right ascension and declination δ_m with a high degree of accuracy and for very little effort. Observations of the Moon and Sun extended over a long period of time should make it possible to predict future values of the right ascension and also their rate of change of right ascension $\dot{\alpha}_m$ and $\dot{\alpha}_s$, and their times of conjunction T.

L is the longitude of the meridian through the sun and Moon at conjunction, and can be calculated as follows. At 0^h the meridian at $180°$ longitude will pass through the Sun, and taking the eclipse of 1981 July 31st as an example, since the time of conjunction $(3^h 35^m.56)$ corresponds to a rotation of the Earth through $53°.89$, the longitude of an observer who sees the eclipse at local noon is approximately $180° + 53°.89 = 233°.89 = -126°.11$. This is approximate because allowance must be made for the difference between mean Sun and true Sun, i.e. for the equation of time. This difference, whose maximum value is !15 minutes is found in this example to be $+6.32$ minutes at the time of conjunction (see Astronomical Almanac Page C13). When this is converted to degrees it has to be subtracted from the approximate longitude calculated above, giving $-127°.69$ as is found on Page A79 of the Almanac. It is possible to establish the equation of time as follows. Assuming that the number of mean solar days in a sidereal year is known, the time of meridian transit of the mean Sun can be calculated for any day, and assuming also that some sort of clock is available (which could be checked or corrected each night), the difference between the expected time of transit and the actual time could be measured. Or, alternatively, the angle could be measured between the true and mean Suns at the time of transit of the mean Sun. This difference is the equation of time.

| Notes |

1. $\hat{AED} = \hat{EDB} \approx$ angle subtended at s by S_t. Since the radius of the Earth is 0.1 arc minutes from the Sun and 58 arc minutes from the Moon, the maximum error here is 0.2%. For the example given the error is 0.1%. This error could be eliminated as $\hat{EDB} = \frac{sp_s}{\cos\delta_s}$ where p_s and δ_s are the Sun's parallax and declination. Then the final equation becomes

$$s = \left[\left(\dot{\alpha}_m - \dot{\alpha}_s\right) + \frac{sp_s}{\cos\delta_s}\right]\frac{\cos\delta_m}{Pm}.$$

2. The maximum possible error in cancelling the sines is 0.001%.

3. The term $\frac{\dot{\alpha}_s t}{60}$ is greatest when t is greatest, the maximum value amounting to an error of 0.1%. This term could be left in the equation.

4. The maximum error due to the oblateness of the Earth (i.e. at the poles) is 0.3%. In this example it is 0.1%.

5. The values given are those are those at conjunction. The maximum error is 0.05%.

6. This is also the value at conjunction, maximum error being 0.04%.

Further notes:

7. The central eclipse begins at $2^h 18^m$ and ends at $5^h 14^m$, so that the example used (at $5^h 00^m$) is almost at the end of the end of the total eclipse phase. Before $5^h 00^m$ the errors involved in using formula (2) are less than 5 sec, and they increase slightly (9sec at $5^h 10^m$) afterwards.

8. The maximum error that could accrue in any eclipse from using formula (2) is estimated to be ± 25 sec.

9. An error also arises in the case of an observer in the umbral shadow but not on the central line. The maximum value of this is ± 12 sec.

Solution of the Eclipse Equation

Solution of the equation $t = 89.544 \sin\left(49.61 + \frac{1}{4}\right)$.

The equation is first modified by changing to the complementary angle as $\left(49.61 + \frac{1}{4}\right) > 45°$.

Therefore, $t = 89.544 \cos\left(40.39 - \frac{1}{4}\right)$.

The factor 0.017453 (equal to $0.02\bar{3}453$) is then inserted to change degrees to radians, and the vinculum is used to remove digits over 5.

We now have $t = 1\bar{1}0.\bar{5}44 \cos\left(0.02\bar{3}453\left(40.4\bar{1} - \frac{1}{4}\right)\right)$ ...(3)

To obtain a first estimate t_0, since $\left(40.4\bar{1} - \frac{1}{4}\right)$ is small and cannot be negative, $\cos\left(40.4\bar{1} - \frac{1}{4}\right)$ is nearly 1, and therefore $t_0 = 80$ is a reasonable value.

The method used here is to insert $t_0 = 80$ into the right-hand side of equation (3) and evaluate it to two significant figures. This gives a second estimate $t_1 = 85$. The calculation continues by adding in the terms arising from the additional digit 5, rather than computing the right-hand side afresh with $t_1 = 85$ as in the usual iterative procedure. A value $t_3 = 84.54$ is obtained after three such cycles, and the method is capable of extension to any number of figures.

Recalling the series expansion: $\cos\theta = 1 - \frac{\theta^2}{2!} + \frac{\theta^4}{4!} - \ldots$, the following operations are performed in sequence on any value entering the right-hand side of equation (3):

a) divide by 4,

b) subtract from $40.4\bar{1}$,

c) multiply by $0.02\bar{3}453$ (θ)

d) square (θ^2)

e) divide by 2 $\left(\frac{\theta^2}{2!}\right)$

f) subtract from 1 $\left(1 - \frac{\theta^2}{2!}\right)$

    (also, where necessary, add $\frac{\theta^4}{4!}$ and higher powers)

g) multiply by $\bar{1}\bar{1}0.\bar{5}44$

The chart is set up as shown in Figure 10 and consists of 8 rows and 3 columns. To refer to any one of the 24 locations a notation will be used in which two numbers, in brackets, will denote the row and column, in that order, of the location. Thus $\bar{5}$ is found in (7,2). The three columns correspond to the three cycles.

		1	2		3		
1		$4 + \bar{2}$	$0 + \bar{1}$	$4 + \bar{1}$	$\bar{1} + \bar{3}$		$\left(40.4\bar{1} - \frac{1}{4}\right)$
2	.0	2	$\bar{3}$	4	5		
3		.3	$_4 4$	$_i \bar{3}$	$_{\bar{5}}$		$0.02\bar{3}453\left(40.4\bar{1} - \frac{1}{4}\right) = \theta$
4		.1 1	$_4 5$	$_0 \overline{17}$	$_{\bar{4}}$		θ^2
5		.0 6	$_i \bar{3}$	$_1 \bar{3}$	$_{\bar{1}}$		$\frac{\theta^2}{2}$
6		$1.\bar{1} 4$	3	8			$\cos\theta = 1 - \frac{\theta^2}{2!} + \frac{\theta^4}{4!} - \ldots$
7		$1\,\bar{1}\,0.$	$\bar{5}$	4	4		
8		$1\,\bar{2}\,5.$	$_1 \bar{6}$	$_0 1\,4$			$t = \bar{1}\bar{1}0.\bar{5}44 \cos\theta$

Figure 10

Rows 2, 7 and the first figure in each column in row 1 are the three constants in equation (3). The other figures in row 1 are from $-\frac{1}{4}$. Thus for $t_0 = 80$, $-\frac{t_0}{4} = \overline{20}$, therefore $\bar{2}$ is put in (1,1) as shown.

Row 3: take the first two cross-products of rows 1 and 2

(i.e. $CP\begin{pmatrix} 2 \\ 2 \end{pmatrix}$ = 4 and $CP\begin{pmatrix} 2 & 0 \\ 2 & \bar{3} \end{pmatrix}$ = $\bar{6}$) to get $4\bar{6}$ = 34, put down 3 carry 4. All carry figures are

written as subscripts.

Row 4: square row 3: the first two duplexes give 114, put down 11 carry 4.

Row 5: $11 \div 2 = 6$ remainder $\bar{1}$ as shown.

Row 6: $1 - 0.06 = 1.\bar{1}4$.

Row 8: the first three cross-products of rows 6 and 7 give $1\bar{2}5$, i.e. $t_1 = 85$.

To start the second cycle the 5 in (8,1) is divided by 4: $5 \div 4 = 1$ rem 1.

The quotient is put, as $\bar{1}$, in (1, 2) and the remainder is carried to (8, 2).

Row 3: since (1, 2) is now modified to $\bar{1}$, row 3 is modified to .32. but as the third CP of rows 1 and 2 gives $19 = 2\bar{1}$, which belongs in (3, 3), the 2 of $2\bar{1}$ is carried leftwards to (3, 2) so that 4 is put in (3, 2) and $\bar{1}$ is carried.

Row 4: take the third duplex of row 3, i.e. 10. This belongs in (4, 3) so that the 1 can be carried leftwards to give 5 in (4, 2) and 0 is carried.

Row 5: combine the carried $\bar{1}$ (as $\overline{10}$) in (5, 2) with 5 in (4, 2) to get $\bar{5}$.

$\bar{5} \div 2 = \bar{3}$ rem 1, put $\bar{3}$ in (5, 2) and carry 1.

Row 6: change the sign of $\bar{3}$ in (5, 2), i.e. put 3 in (6, 2).

Row 8: take the fourth cross-product of rows 6 and 7 to get $\bar{6}$, put in (8, 2), i.e. t = 84.4.

To start the third cycle combine $\bar{6}$ in (8, 2) with the carried 1 to get 4.

$4 \div 4 = 1$ rem 0, put the quotient, as $\bar{1}$, in (1, 3) and carry 0 to (8, 3).

The beauty of this method is that it is totally efficient. No unnecessary work is done, only those figures needed are calculated and the remainders allow further digits of the answer to be produced later, if required.

Chapter 3

KEPLER'S EQUATION

Kepler's Equation is of great importance in positional astronomy. It is used to get the position of a planet in its elliptical orbit at a given time.

The equation is: $$M = E - e \sin E$$

where M and E are the mean and eccentric anomalies respectively of a planet in its orbit (in radians) and e is the eccentricity of the orbit.

The eccentricity, e, is a measure of the oblateness of the ellipse and can be defined by $e^2 = 1 - \frac{b^2}{a^2}$ where a and b are the lengths of the semi-major and semi-minor axes as shown below.

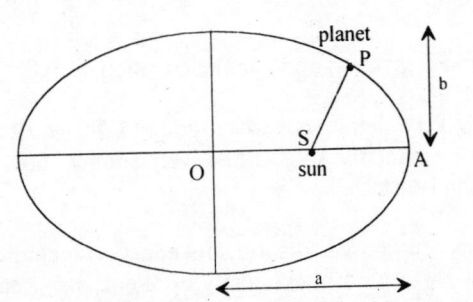

Figure 1: The elliptical orbit of a planet P around the Sun S.

The velocity of a planet in an elliptical orbit is variable and the mean anomaly, M, is the angle shown in the diagram below assuming the planet travelled at constant speed in a circular orbit that was completed in the same time as the actual planet, and started at A at the same instant.

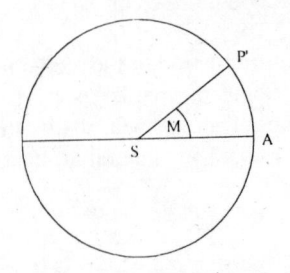

Figure 2: The Mean Anomaly M.

M is found from $\dfrac{M}{2\pi} = \dfrac{d}{T}$ where d is the number of days since the planet passed A (called the perihelion) and T is the time for one full orbit.

The eccentric anomaly, E, is as shown in the diagram below where Q lies on a circle of radius a. v is the true anomaly of the planet at P and there is a fairly simple formula for getting v once E is known.

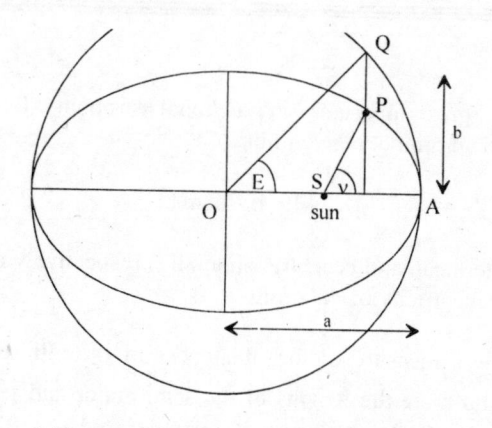

Figure 3: The eccentric anomaly E and the true anomaly v.

So the problem is to find E given M and e in the equation $M = E - e\sin E$

Kepler's equation looks very simple, but as we require E the equation is a transcendental one, which means E cannot be made the subject of the equation without having an infinite number of terms on the right-hand side.

Many methods have been proposed to solve this equation including one by Kepler himself. The Vedic method which follows is extremely efficient, using each digit of the answer as they are obtained to get the next digit. Modern calculating devices make the solution rapid, using the Newton-Raphson, or some other, iterative technique. But their methods, though extremely fast are not efficient and there is also a need for quick pencil and paper solutions.

Before an example of the solution of Kepler's equation a simpler transcendental equation will be solved. This will show the method more clearly.

Chapter 1 of this book gives an introduction to the Vedic methods which enable efficient left to right calculations to be made and the reader is advised to look at that material if not familiar with the Vedic techniques. That is, specifically, the use of cross-products, duplexes, left to right calculations and the use of the vinculum (negative digits).

3.1 A TRANSCENDENTAL EQUATION

Solve the transcendental equation $x + \sin x = 1$

Using the series expansion for sinx: $x + x - \dfrac{x^3}{3!} + \dfrac{x^5}{5!} - \ldots = 1$

$$\therefore x = 0.5 + \frac{x^3}{12} - \frac{x^5}{240} + \frac{x^7}{2\times7!} - \ldots \qquad\qquad \ldots (1)$$

Since x is less than 1 the terms on the right of this equation decrease in value. So we use 0.5 as an initial estimate for x. Then we find the first figure of x^3 and divide this by 12, noting the quotient and remainder. This gives us the next significant figure of x and we compute the next figure of $\frac{x^3}{12}$ and so on.

$$
\begin{array}{rl|ll}
\therefore & 0.5 & 0 \;.\; 5 & \text{row 1} \\[4pt]
& \frac{x^3}{12}\quad 12 & & \text{row 2} \\[4pt]
& -\frac{x^5}{240}\quad 2 & & \text{row 3} \\[2pt]
\hline
& \mathbf{x} = & \mathbf{0 \;.\; 5} & \text{row 4} \\[4pt]
& (x^2 = & & \text{row 5}
\end{array}
$$

We begin by setting up the chart above. On the left of the vertical line we have the terms on the right-hand side of equation (1) above. The subscripts 12 and 2 are the divisors for that row. The 12 is obviously a divisor in row 2. The value of x^2 is formed in row 5 and this is needed to get row 3 from row 2 by multiplying row 2 by row 5. But since x^5 is divided by 240 a further division of 20 is necessary. That is why 2 is put as a divisor in row 3. Row 5 can also be used to form the value in the x^7 row, if it was needed, by multiplying row 3 by row 5 and dividing by the appropriate number.

The first step is to put 0.5 over to the right of the vertical line and then bring it down into the answer in row 4, as shown above.

Next we get the first figure of x^2 by squaring row 4.

The duplex of 5, D(5) is 25 and we put down 2 carry 5, written as 2_5 or 3 carry $\bar{5}$, written as $3_{\bar{5}}$, in this case the latter (see below).

Having got the first figure of x and x^2 we multiply these to get the first figure of x^3, 15, and this must be divided by 12.

This gives 1 carry 3 as can be seen below in row 2. Next we bring down this 1 into the answer.

$\therefore$

	0.5		0 . 5								row 1
	$\frac{x^3}{12}$	12		1_3							row 2
	$-\frac{x^5}{240}$	2									row 3
	x =		0 . 5 . 1								row 4
	(x² =		. 3_5								row 5

Now we can get the next figure of x^2: looking at row 4, D(51) = 10, and this combined with the carried $\bar{5}$, which is $\overline{50}$ in the second decimal place, gives $\overline{40}$ so we put down $\bar{4}_0$ in row 5.

$\therefore$

	0.5		0 . 5								row 1
	$\frac{x^3}{12}$	12			1_3 1_1 1_2 2 $\bar{2}_2$						row 2
	$-\frac{x^5}{240}$	2				$\bar{1}_1$ $\bar{5}_1$					row 3
	x =		0 . 5	1 1 0 3							row 4
	(x² =		. 3_5 $\bar{4}_0$ 1_1 1_2 $\bar{1}_1$)								row 5

Now we can find the next figure of x^3 by finding the next cross-product of rows 4 and 5.

This is $CP\begin{pmatrix} 5 & 1 \\ 3 & \bar{4} \end{pmatrix} = \overline{20} + 3 = \overline{17}$.

Now looking at row 2 there is a carry of 3, so we add 30 to the $\overline{17}$ which gives 13, and this we divide by 12 to get 1 carry 1. So we put down 1_1 in row 2.

Bring down 1 from row 2 into the answer.

Then in row 4, D(511) = 11, put 1_1 in row 5.

Then $CP\begin{pmatrix} 5 & 1 & 1 \\ 3 & \bar{4} & 1 \end{pmatrix} = 4$ (this is for x^3). This 4 + carried 1 in row 2 gives 14 and $14 \div 12 = 1_2$

which we put down in row 2.

Now we have ignored row 3 and this comes in next. Row 3 comes from multiplying rows 2 and 5 and dividing by 20. A little thought will show that this gives its first significant figure in the 4th decimal place.

The first cross-product of rows 2 and 5 is $CP\begin{pmatrix} 1 \\ 3 \end{pmatrix} = 3$.

Dividing this by 2 we get 1_1 and because the term being evaluated is negative we put down $\bar{1}_1$ in row 3.

Now we can bring 0 down into the answer.

Then row 4: D(5110) = 2.

Row 5: 2 + 10 = $1_{\bar{2}}$ put it down.

Rows 4,5: $CP\begin{pmatrix} 5 & 1 & 1 & 0 \\ 3 & \bar{4} & 1 & 1 \end{pmatrix} = 2.$

Row 2: 2 + 20 = 22, 22÷12 = $2_{\bar{2}}$ put it down.

Rows 2,5: $CP\begin{pmatrix} 1 & 1 \\ 3 & \bar{4} \end{pmatrix} = \bar{1}.$

Row 3: $\bar{1}$ + 10 = 9, 9÷2 = 5_i put it down.

Bring $\bar{3}$ down into the answer.
And so on.

The method is quick and efficient if x is small and can be used to extend the solution to any number of decimal places.

Using the figures of the answer to get the next answer figure, the procedure is similar to that in iterative techniques. But only those figures actually required to get the next figure are calculated so that not a single superfluous step is taken: the method is one hundred per cent efficient. And with the extra flexibility offered by the vinculum device the figures can always be kept small and manageable.

3.2 SOLUTION OF KEPLER'S EQUATION

Now let us look at Kepler's Equation. We will take M = 0.30303 radians and e = 0.016722, which is the eccentricity of the Earth's orbit.

Solve, 0.30303 = E − 0.016722 sinE

$$\therefore E = 0.30303 + 0.016722\left(E - \tfrac{E^3}{3!} + \tfrac{E^5}{5!} - ....\right) \qquad \qquad ...(2)$$

M	0 . 3 0 3 0 3 0 0	1
eE	1 $_{\bar{4}}$ $\bar{5}$ 3 2 $_{\bar{6}}$ $\bar{5}$ $_1$ 2 $_1$ 1 $_{\bar{3}}$	2
$-\tfrac{eE^3}{3!}$ 6	$\bar{1}$ $_i$ 2 $_i$ $\bar{1}$ $_3$ $\bar{5}$ $_0$	3
$\tfrac{eE^5}{5!}$ 2	4 $_0$	4
E =	0 . 3 1 $\bar{2}$ 1 0 1 0	5
(E² =	. 1 $_i$ 0 $_{\bar{4}}$ $\bar{5}$ $_i$ $\bar{1}$ 2 3 $_{\bar{4}}$)	6

where $e = 0.016722 = 0.02\overline{3}\overline{3}22$

We use $6 = 3!$ as a divisor in row 3 and for the same reason as in the previous example we use 2 as a divisor in row 4.

Bring 0.3 into the answer row.

Row 2: we multiply e and E, $CP\begin{pmatrix} 2 \\ 3 \end{pmatrix} = 6 = 1_4$

Bring 1 into the answer row.

Row 2: $CP\begin{pmatrix} 2 & \overline{3} \\ 3 & 1 \end{pmatrix} = \overline{7}, \ \overline{7} + \overline{40} = \overline{5}_3$

Bring $\overline{2}$ down into the answer row.

Row 2: $CP\begin{pmatrix} 2 & \overline{3} & \overline{3} \\ 3 & 1 & \overline{2} \end{pmatrix} = \overline{16}, \ \overline{16} + 30 = 2_{\overline{6}}$

Row 3: we find E^2 (row 6) then multiply rows 2 and 6, divide by 6 and change the sign:

$$CP\begin{pmatrix} 1 \\ 1 \end{pmatrix} = 1, \ CP\begin{pmatrix} 1 & \overline{5} \\ 1 & 0 \end{pmatrix} = \overline{5}, \ \text{giving } 1\overline{5} = 5,$$

$5 \div 6 = 1 \, r\overline{1}$, put $\overline{1}_1$

Bring 1 down into the answer.
Etc.

We get $\underline{E = 0.3081010}$ to 7 decimal places.

Such accuracy is superfluous in this case as $e = 0.016722$ is only known to 5 significant figures.

Although M, and therefore E, take all values between 0 and 2π, by using the angle to the nearest quadrant boundary we can always arrange that the quantity we calculate is less than $\frac{\pi}{4}$.

For example, if M=3 radians, so that $E \approx 3$, we let $E = \pi - A$

$\therefore M = \pi - A - e\sin(\pi - A)$
$\quad = \pi - A - e\sin A$

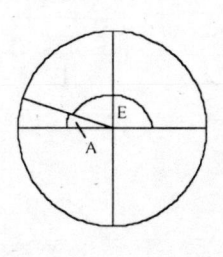

from which we can find A and then E, since $E = \pi - A$.

Another Example

We will find E for the Earth 118 days after perihelion passage.

Then $M = \dfrac{2\pi \times 118}{365.25636} = 2.0298507$ and $e = 0.016722$.

The Earth is then just after the 90° quadrant boundary. We therefore put $E = \dfrac{\pi}{2} + A$.

So $M = E - e\sin E$ becomes $M = \dfrac{\pi}{2} + A - e\sin(\dfrac{\pi}{2} + A)$

Or $M - \dfrac{\pi}{2} = A - e\cos A$.

We will therefore find A and then E from $E = \dfrac{\pi}{2} + A$.

So $A = M - \dfrac{\pi}{2} + e - \dfrac{eA^2}{2!} + \dfrac{eA^4}{4!} - \ldots$

That is: $A = 0.52422434 - 0.02\bar{3}322\,\dfrac{A^2}{2} + 0.02\bar{3}322\,\dfrac{A^4}{24} - \ldots$ ... (3)

$M - \dfrac{\pi}{2} + e$		0 .	5	$\bar{2}$	$\bar{4}$	$\bar{2}$	2	$\bar{4}$	3	4		1
$-\dfrac{eA^2}{2}$	2				$\bar{2}$	0_1	2 $_i$	5 $_0$	$\bar{3}1$ $_0$			2
$\dfrac{eA^4}{24}$	12						3 $_8$	5 $_3$	1 $_{\bar{3}}$			3
$A =$		0 .	5	$\bar{2}$	6	$\bar{1}$	3	3	3			4
($A^2 =$		.	2 $_5$	3 $_0$	5 $_{\bar{6}}$	5 $_2$	11 $_0$)					5

First bring 5 down into the answer. Square it (for row 5) and put down 2_5.

To get $-\dfrac{eA^2}{2}$ in row 2 multiply the first figure in row 5 by the first figure of $e = 0.02\bar{3}322$: $2 \times 2 = 4$, and divide this by 2: $4 \div 2 = 2$. So put $\bar{2}$ in row 2.

Bring $\bar{2}$ down into the answer. Then $D(5\bar{2}) = \overline{20}$. There is a 5 carried in row 5 so $\overline{20} + 50 = 30$: put 3_0.

We next need the next figure in row 2 so we again multiply row 5 by e: $CP\begin{pmatrix} 2 & \bar{3} \\ 2 & 3 \end{pmatrix} = 0$, $\div 2 = 0$, put 0 in row 2.

Add the third column and put $\overline{6}$ in the answer row. The next duplex is $D(5\,\overline{2}6)=\overline{5}6$ (and nothing carried) so put down $\overline{5}_6$ in row 5.

Next $CP\begin{pmatrix} 2 & \overline{3} & \overline{3} \\ 2 & 3 & \overline{5} \end{pmatrix}=\overline{25},\ \overline{25}\div 2=\overline{1}2_1$ put 12_1 in row 2.

Consider next the A^4 term. Row 2 will be multiplied by row 5 and the result divided by 12. The sign must also be changed.

Therefore finding the first two duplexes of rows 2 and 5: $CP\begin{pmatrix} \overline{2} \\ 2 \end{pmatrix}=\overline{4}$ and $CP\begin{pmatrix} \overline{2} & 1 \\ 2 & 3 \end{pmatrix}=\overline{4}$. So

we have $\overline{44}$, which we change to 44. The $44\div 12=3_8$ which we put in row 3.

And so on.

Finally, adding the $\dfrac{\pi}{2}$ back on, since $E = \dfrac{\pi}{2} + A$, we get $E = 2.0447296$.

Such accuracy is unnecessary, but it is possible. If the answer were required to 4 decimal places, which is equivalent to about ± 10 arc seconds the calculation becomes:

$$
\begin{array}{r|ccccc}
 & 0\ .\ 5 & \overline{2} & 4 & \overline{2} \\
 & & & \overline{2} & 0_1 \\
\hline
A = & 0\ .\ 5 & \overline{2} & 6 & \overline{1} \\
 & .\ 2 & {}_5\,3 & {}_0\,\overline{5} & {}_{\overline{6}}
\end{array}
$$

Giving $E = 2.0447$.

This method has many other applications including the calculation of sines, cosines etc. and their inverses, and hyperbolic and polynomial equations. See "Vertically and Crosswise".

Chapter 4

INTRODUCTION TO TRIPLES

A Pythagorean triple is a set of three real numbers x, y, r such that $x^2 + y^2 = r^2$.

For example

                                    3, 4, 5
                                    12, 5, 13
                                    etc.

From Pythagoras' Theorem these triples can therefore represent the sides of a right-angled triangle.

Pythagorean triples have been known and used for thousands of years; long before Pythagoras in fact. Early Babylonians (1900-1600 B.C.) recorded some triples on a stone tablet that is now part of the Plimpton Collection at Columbia University, New York.

The number of distinct triples is infinite, even if we restrict the three elements of a triple to be natural numbers and count all multiples of a triple as equal. A triple of any desired shape and size can be found to any desired degree of accuracy.

4.1 NOTATION AND COMBINATION

Let us suppose that the first element of a triple is the base of a right-angled triangle, the second is the height and the third is the hypotenuse. And also that the angle in a triple is the angle between the base and hypotenuse (so the base is not necessarily at the bottom).

If the angle in the triple is A, then we can write A) 4, 3, 5 to denote the triangle below.

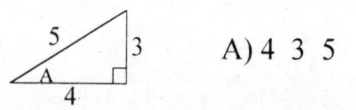

                                       A) 4 3 5

This compact notation is very comprehensive, including both the polar method of describing the position of a point (the outer pair, 5, A, above) and the Cartesian method (the inner pair, 4, 3, above).

Since the triangle for the triple px, py, pr (where p is a real number) is the same shape as that for x, y, r we will consider these to be **equal triples**. So we might say for example A) 4, 3, 5 = A)12, 9, 15 = A)2, 1½, 2½.

Further it is useful to call triples in which all the elements are rational, **perfect triples**.
So 2, 1½, 2½ is a perfect triple but 3, 1, $\sqrt{10}$ is not.

Triple Addition

Next we need to define a way of combining triples.

Suppose we have two triples:

$$A) \ 4 \quad 3 \quad 5$$
$$B) \ 15 \quad 8 \quad 17$$

and that we wish to add them in the way shown below, to produce a triangle OQP containing
the angle (A + B).

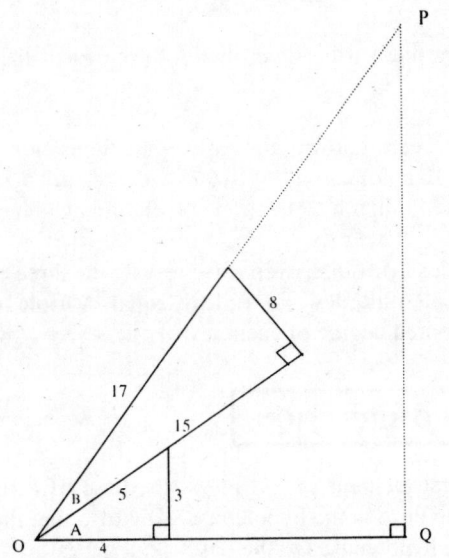

A triple which contains the angle (A + B) is given by:

A |	4	3	5	
B |	15	8	17	+
A+B |	(4×15 – 3×8),	(3×15 + 4×8),	(5×17)	
= |	36	77	85	

That is, we multiply vertically in the first two columns and subtract to get the first element of
the triple for (A + B).

Then we cross-multiply in the first two columns and add to get the second element.

And we multiply vertically in the third column to get the third element.

There is a simple Vertical and Cross-wise pattern here which is shown diagrammatically below:

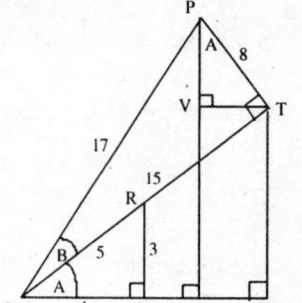

1st element   2nd element   3rd element

In general, for two triples A) x   y   r   and B) X   Y   R   the triple containing the angle (A + B) is given by:

$$
\begin{array}{c|ccc}
A & x & y & r \\
B & X & Y & R \\
\hline
A+B & (xX - yY), & (yX + xY), & (rR)
\end{array} \quad + \qquad \qquad \ldots(1)
$$

This can be demonstrated as follows, using the triples in the above example.

We require the sides of triangle OQP, which contains the angle A+B.

Comparing the similar triangles OSR, PVT:
  OS is $\frac{4}{5}$ of the hypotenuse OR, so PV = $\frac{4}{5}$ of 8.
  and RS = $\frac{3}{5}$ of the hypotenuse, so TV = $\frac{3}{5} \times$ 8.

Similarly, comparing triangles OSR, OUT:
  OU = $\frac{4}{5} \times$ 15,
  TU = $\frac{3}{5} \times$ 15.

Now, in triangle OQP: OQ = OU – UQ = OU – TV = $\frac{4}{5} \times$ 15 – $\frac{3}{5} \times$ 8,
        and   QP = VQ + PV = TU + VP = $\frac{3}{5} \times$ 15 + $\frac{4}{5} \times$ 8.

Therefore a triple for OQP is: $\frac{4}{5} \times$ 15 – $\frac{3}{5} \times$ 8, $\frac{3}{5} \times$ 15 + $\frac{4}{5} \times$ 8,   17

                = 4 × 15 – 3×8,   3 × 15 + 4 × 8,   5 × 17 (on multiplying
                                                                                through by 5)
And this is just the pattern of products we used to add the triples.

Replacing the numbers 4, 3, 5 and 15, 8, 17 with x, y, r and X, Y, R gives the general result above.

Thus it turns out that the sum of two perfect triples is itself a perfect triple!

Also if A) x, y, r **the double angle triple** is given by 2A) x² – y², 2xy, x² + y²   . . . (2)
(this follows directly from (1) above by putting A = B in the triple for (A + B)).

Quadrant Triples

The quadrant angles, 0°, 90°, 180°, 270° can also be expressed by triples.

Suppose we add the triples A)4 3 5 and B)3 4 5:

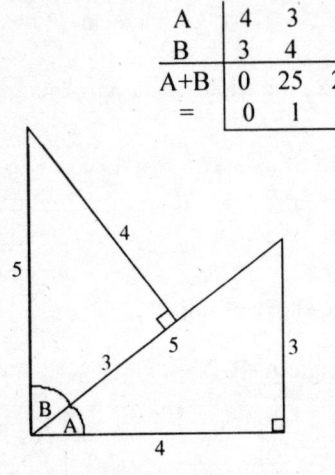

A	4	3	5
B	3	4	5 +
A+B	0	25	25
=	0	1	1

In this case we get a triangle with no base and height and hypotenuse equal.

We can divide 0, 25, 25 by 25 to get 0, 1, 1.

We see that because the triples are complementary A + B = 90°, so that the triple 0, 1, 1 represents an angle of 90°:

90°) 0   1   1

And if 90°) 0 1 1 we can double this triple to get a triple for an angle of 180°:

90°	0	1	1
180°	−1	0	1

Next we can double the 180° triple to get a triple for 360°, which is also a triple for 0°:

180°	−1	0	1
360°	1	0	1

And for 270° we can add 180° and 90°:

180°	−1	0	1
90°	0	1	1
270°	0	−1	1

These triples are summarised below:

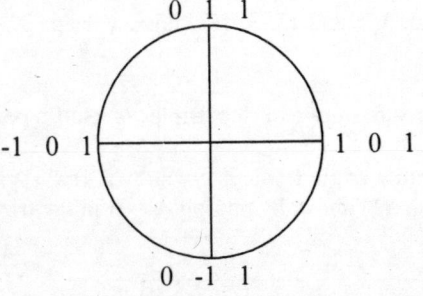

Rotations

This simple combination process leads to huge numbers of applications.
Suppose we want to rotate the point P(5, 2) through the angle in the triple A)4, 3, 5.

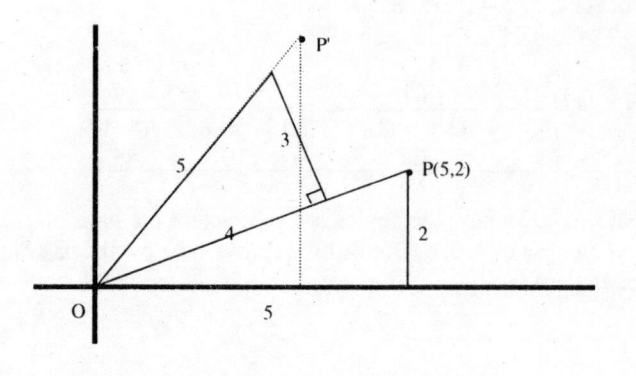

The point (5, 2) gives us a right-angled triangle: 5, 2, $\sqrt{29}$.
We want the coordinates of P'.

As can be seen from the diagram, rotating OP through A) 4,3,5 is equivalent to adding these triples.

$$
\begin{array}{ccc}
5 & 2 & \sqrt{29} \\
4 & 3 & 5 \\
\hline
14 & 23 & 5\sqrt{29}
\end{array} \quad +
$$

But we want the hypotenuse of the resulting triangle to be equal to OP, which is $\sqrt{29}$.
We therefore divide through by 5 which gives P'$(\frac{14}{5}, \frac{23}{5})$ or P'(2.8, 4.6).

Note that the third column in the calculation above is unnecessary as we consistently divide through by the third element of the rotation triple.

This technique is perfectly general; the rotation triple does not have to be a perfect triple, we could rotate through the angle in the triple 3, 1, $\sqrt{10}$ for example, if we chose.

One advantage of the triple method is that we avoid dealing with the angle, which is generally incommensurable with the sides of the triangle.

Triple Subtraction

This is the same as for triple addition except that we add where before we subtracted and subtract where before we added.

For example, for A) 4, 3, 5 – B) 15, 8, 17:

A |	4	3	5	
B |	15	8	17	–
A–B |	(4×15 + 3×8),	(3×15 – 4×8),	(5×17)	
= |	84	13	85	

The diagram below shows how the angles are subtracted: the base of 15, 8, 17 is placed as before on the hypotenuse of 4, 3, 5. But its hypotenuse is below the base so that the angle B is subtracted from angle A.

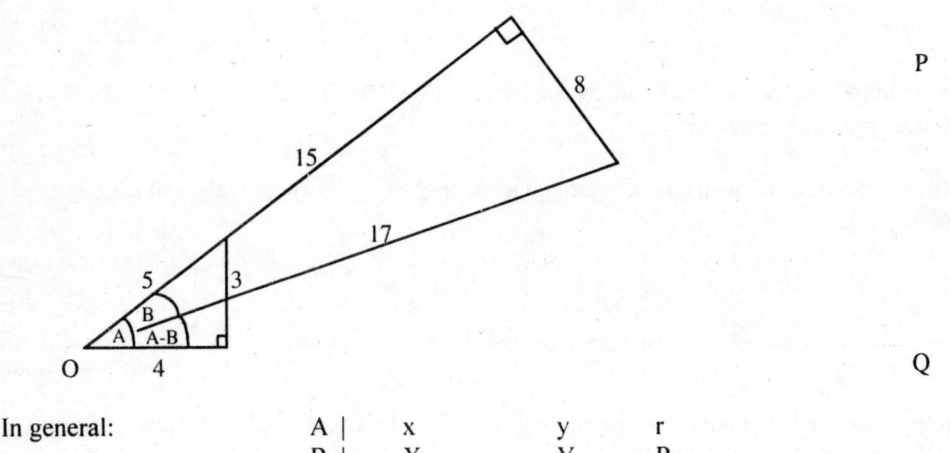

In general:

A |	x	y	r	
B |	X	Y	R	–
A–B |	(xX + yY),	(yX – xY),	(rR)	. . . (3)

Note: where one element of a triple is unknown or unwanted we may put a dash (see below).

The Half-Angle Triple

Given A) 7   24   25 the triple for ½A is

$$\begin{aligned} ½A)\ &7{+}25,\ 24,\ \text{---} \\ =\ &32,\ 24,\ \text{---} \\ =\ &4,\ 3,\ \text{---} \\ =\ &\underline{4,\ 3,\ 5} \end{aligned}$$

That is, we add the first and last elements of the triple to get the first element of the half-angle triple and keep the middle element as the middle element of the half-angle triple.

The third element can be calculated from the first two after any cancelling has been carried out.

This is easily demonstrated as follows.

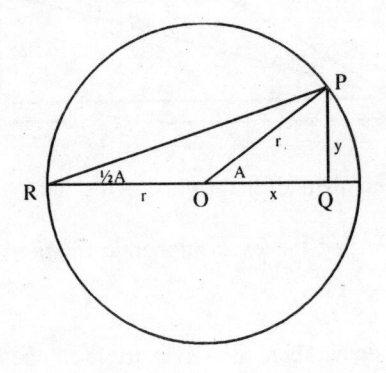

Given a triple A) x, y, r for the triangle OQP draw a circle centre O, radius r, as shown. Extend QO to meet the circle at R and join RP.

Then the angle at R is ½A so that from ΔRQP the triple for ½A is ½A) x + r, y, -.

So, given A) x, y, r then ½A) x + r, y, -.   ...(4)

And given A) 4, 3, 5 then ½A) 4+5, 3, - = 3, 1, $\sqrt{10}$.

4.2   TRIPLE CODE NUMBERS

For all integer values of c, d

$c^2 - d^2$, 2cd, $c^2 + d^2$ generates perfect triples.   ...(5)

This is shown by adding the squares of the first two elements of $c^2 - d^2$, 2cd, $c^2 + d^2$ and showing that the result is the same as squaring the third element. In fact it can be shown that all perfect triples are generated by this formula.

Since $c^2 - d^2$, 2cd, $c^2 + d^2$ is also the result of doubling the triple c,d,- (see equation (2)) it follows that the triple generated by values c, d has twice the angle in the triple $c^2 - d^2$, 2cd, $c^2 + d^2$, when c, d is considered as a triple.

i.e. diagrammatically:

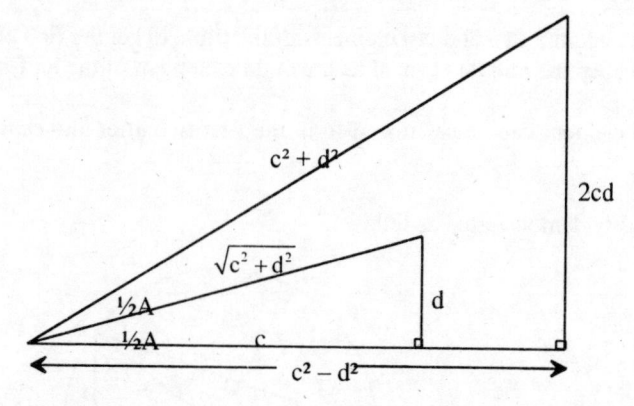

c, d will be called the code numbers of the triple $c^2 - d^2$, $2cd$, $c^2 + d^2$.

We can see from the above that the exact algebraic relationship between A and its code numbers is $\frac{c}{d} = \cot \frac{1}{2} A$.

It further follows that the code numbers of a given triple can be found by using the method of finding a half-angle triple.

That is, the code numbers of x,y,r are x+r,y,-.   ...(6)

So given the triple 4,3,5 the half-angle triple is 4+5, 3, - = 9, 3, - = 3, 1, -
so c=3, d=1 generate the triple 4,3,5.

When d=1 in a code number pair we can omit this value and simply write c=3.
So a single value for a code number means that d=1: if we write c=3 we infer d=1.

So we have:

4	3	5	c=3
3	4	5	2
–3	4	5	$\frac{1}{2}$
–4	3	5	$\frac{1}{3}$
–4	–3	5	$-\frac{1}{2}$
–3	–4	5	$-\frac{1}{3}$
3	–4	5	–2
4	–3	5	–3

We can plot these triangles and their code numbers, together with those for the quadrant boundaries (and 4 other triangles), on a single diagram:

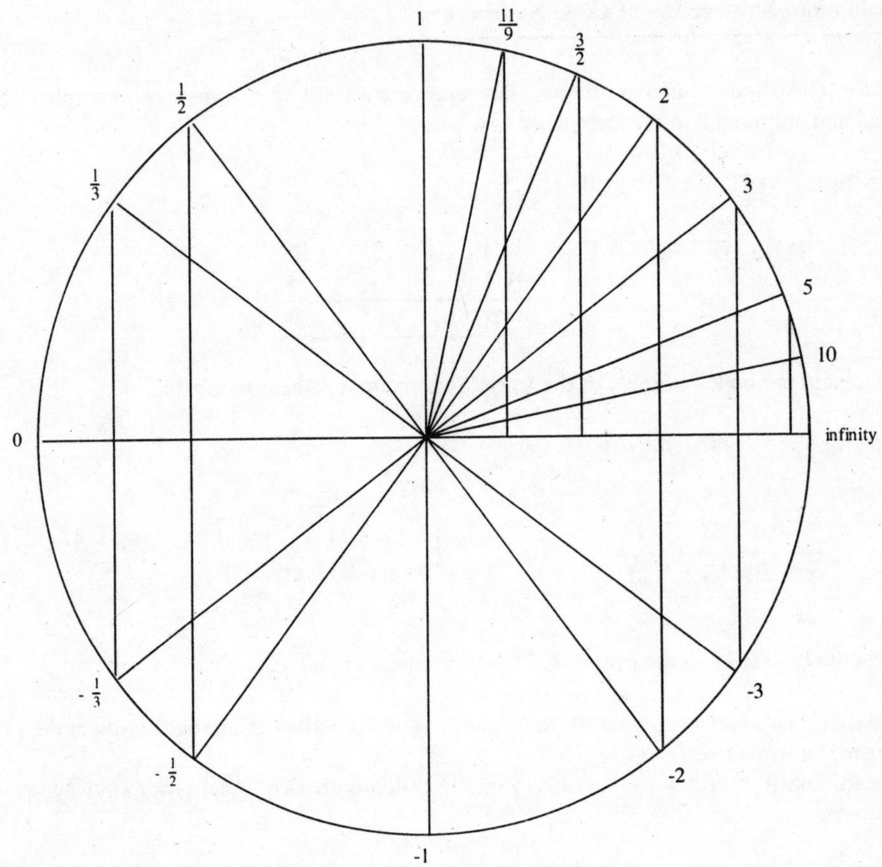

(diagram of circle with radial lines labeled with code numbers: 1, $\frac{11}{9}$, $\frac{3}{2}$, 2, 3, 5, 10, $\frac{1}{2}$, $\frac{1}{3}$, 0, infinity, $-\frac{1}{3}$, $-\frac{1}{2}$, -1, -2, -3)

We can see how the code numbers change as the angle increases from 0° to 360° from the above diagram.

As the angle in the triple increases from 0° to 360° the code numbers run from $+\infty$ to $-\infty$.

There is symmetry about the horizontal line. I.e. the code numbers below are minus those above. So $c(-A) = -c(A)$.

There is symmetry about the vertical line. I.e. the code numbers on the left are the reciprocal of those on the right. So $c(180° - A) = \frac{1}{c(A)}$.

Note: If, instead of seeing the code number triangle as having half the angle of the triple it generates, we see the triple as having double the angle of its code number triangle, we have the following useful relationship:

the first two elements of a triple are the code numbers of a triple with double the angle.

That is, for the triple A) x, y, z $c(2A) = \frac{x}{y}$.

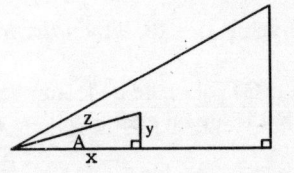

This result will be useful later on.

Addition and Subtraction of Code Numbers

Because a pair of code numbers themselves represent a triangle or triple, code numbers can be added and subtracted in the same way as triples.

Add the triples A) 21, 20, 29 and B) 4, 3, 5.

By the usual method:

A	21	20	29
B	4	3	5 +
A+B	24	143	145

Or we can add the code numbers of these triples and convert them to a triple.

The code numbers for the above triples are 5, 2 and 3, 1:

c(A)	5	2
c(B)	3	1 +
c(A+B)	13	11

where $13 = 5 \times 3 - 2 \times 1$ and $11 = 2 \times 3 + 5 \times 1$
I.e. $c(A) + c(B) = c(A+B)$.

We then convert 13, 11 to the triple 24, 143, 145 using (5) above.

The arithmetic is easier here because the code numbers are always smaller numbers than the numbers in the triple itself.
This is particularly useful when we only want the code numbers of the answer anyway, as we will see later.

The code number sum above can be written as 5, 2 $\dot{+}$ 3, 1 = 13, 11

$$\text{or as } \tfrac{5}{2} + 3 = \tfrac{13}{11}.$$

In order not to confuse triple addition with ordinary addition of numbers we will use a special symbol for triple addition: a plus sign with a dot over it ($\dot{+}$).

So $\tfrac{5}{2} \dot{+} 3 = \tfrac{13}{11}$

Similarly we can use a minus sign with a dot over it ($\dot{-}$) for triple subtraction.

Complementary Triples

Complementary triples add up to 90° and we transpose the first two elements of a triple to obtain its complementary triple.

So if A) x,y,r then 90°−A) y,x,r.

Given c, d = 20, 9 then *By Addition and By Subtraction*

c, d(CT) (i.e. the code numbers of the complementary triple) = 20+9, 20−9 = **29, 11**.
That is, given code numbers c,d then c,d(CT) = c+d,c−d,−

And if c, d = 29, 11 then c, d(CT) = 29 + 11, 29 − 11 = 40, 18 = **20, 9**.

Proof: For the triple $c^2 - d^2$, $2cd$, $c^2 + d^2$ with code numbers c and d the complementary triple is $2cd$, $c^2 - d^2$, $c^2 + d^2$.

Then by the method used above
$c, d(CT) = 2cd + c^2 + d^2, c^2 - d^2 = (c + d)^2, c^2 - d^2 = (c + d),(c - d)$.

4.3   ANGLES IN PERFECT TRIPLES

The table below shows the triples and their code numbers for c = 1 to c = 30.

The angle contained in each triple is also shown, in radians, to 4 decimal places.

The first entry in the table is the triple for 90°, or $\frac{\pi}{2}$ radians.

There are many interesting relationships between the triples in this table.

For example we can see that since 4 + 13 = 3 then 0.4900 + 0.1535 = 0.6435 radians.
It is important at this stage to appreciate that these two statements, one in code numbers and the other in angles, are equivalent (triple addition adds the angles, triple subtraction subtracts the angles).

T R I P L E			ANGLE		T R I P L E			ANGLE	
code no.			A		code no.			A	
c			in rad.	2/c − A	c			in rad.	2/c − A
0	1	1	1.5708		255	16 × 2	257	0.1248	0.0002
3	2 × 2	5	0.9273		144	17	145	0.1175	0.0001
4	3	5	0.6435	0.0232	323	18 × 2	325	0.1110	0.0001
15	4 × 2	17	0.4900	0.0100	180	19	181	0.1052	0.0001
12	5	13	0.3948	0.0052	399	20 × 2	401	0.0999	0.0001
35	6 × 2	37	0.3303	0.0030	220	21	221	0.0952	0.0001
24	7	25	0.2838	0.0019	483	22 × 2	485	0.0908	0.0001
63	8 × 2	65	0.2487	0.0013	264	23	265	0.0869	0.0001
40	9	41	0.2213	0.0009	575	24 × 2	577	0.0833	0
99	10 × 2	101	0.1993	0.0007	312	25	313	0.0800	0
60	11	61	0.1813	0.0005	675	26 × 2	677	0.0769	0
143	12 × 2	145	0.1663	0.0004	364	27	365	0.0740	0
84	13	85	0.1535	0.0003	783	28 × 2	785	0.0714	0
195	14 × 2	197	0.1426	0.0002	420	29	421	0.0689	0
112	15	113	0.1331	0.0002	899	30 × 2	901	0.0666	0

We may observe from this table that there is a simple relationship between the code numbers and their angles:

> the product c.A is approximately equal to 2,

and gets closer to 2 further down the table, i.e. as c increases.
The difference between $\frac{2}{c}$ and A is given in the right-hand column.

This enables us to convert triples or code numbers to angles, or angles to code numbers or triples, given small angles.

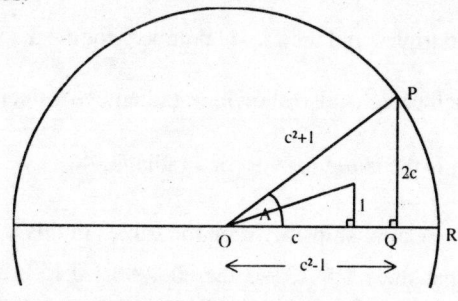

Using the diagram above we can easily prove this result.
If A is measured in radians, the arc PR = $A(c^2 + 1)$.
$\therefore$ as $A \to 0$, $A(c^2 + 1) \to 2c$
$\therefore$ $cA + \frac{A}{c} \to 2$
But as $A \to 0$, $c \to \infty$ $\therefore$ $cA \to 2$.

EXAMPLE 1   Find the triple which contains an angle of 0.04 radians.

Since $c \approx \frac{2}{A}$, $c \approx \frac{2}{0.04} = 50$ which is the code number of the required triple.

So the triple is   <u>2499, 100, 2501</u>.

EXAMPLE 2   Find the angle in the triple   1599, 80, 1601.

Since the code number is 40, the angle is $\frac{2}{40}$ = <u>0.05 radians</u>.

We discuss accuracy later (page 160), but for now we may note that the angles calculated in this way are accurate:

to 2 decimal places for $c \geq 6$,
to 3 decimal places for $c \geq 13$,
to 4 decimal places for $c \geq 24$,
to 5 decimal places for $c \geq 52$.

EXAMPLE 3   Find the angle in the triple   1512, 55, 1513.

$c = 55$ so $A = \frac{2}{55} = \frac{4}{110} = $ <u>0.03636</u>   to 5 decimal places.

4.4 GENERAL ANGLES

Let A be the angle in a triple and c its code number, and let A' and c' be the angle and code number of a perfect triple in which $A \approx A'$
Also let $a = A - A'$, so that a is small.

So $A = A' + a$
and so $c(A) = c(A') + c(a)$.

Or $c = c' + c(a)$
So $c \approx c' + \frac{2}{a}$ and adding these:

$$
\begin{array}{cc}
c' & 1 \\
\dfrac{2}{2c'-a,} & \dfrac{a}{c'a+2} \\
\end{array} +
$$

So $c \approx \frac{2c'-a}{c'a+2}$ ... (7)

which leads to $a \approx \frac{2(c'-c)}{c'c+1}$

so that $A \approx A' + \frac{2(c'-c)}{c'c+1}$ ... (8)

Formulas 7 and 8 allow us to get c given A or A given c for a particular triple by using a nearby perfect triple with angle A' and code number c'.

The following ten examples illustrate the application of these formulae and the results are used in Chapter 5.

EXAMPLE 4   Find c(0.4112).

Here we require the code number of the triple with angle 0.4112 radians. So we use (1) above.

A=0.4112 is close to A'=0.3948, for which c' = 5.

So we have $c(0.4112) = \frac{2(c'-c)}{c'c+1} = \frac{2 \times 5 - 0.0164}{2 + 5 \times 0.0164} = \frac{9.9836}{2.082} = \textbf{4.7952}$.

EXAMPLE 5   Find c(0.7082).

We know c(0.6435) = 3, so c'=3 and a=0.0647.

$\therefore c(0.7082) = \frac{2 \times 3 - 0.0647}{2 + 3 \times 0.0647} = \frac{5.9353}{2.1941} = \textbf{2.7051}$.

EXAMPLE 6   Find c(1.5414).

Here we can use c'=1 and a is therefore 1–5414 – 1.5708 = –0.0294.

$$c(1.5414) = \frac{2 \times 1 + 0.0294}{2 - 1 \times 0.0294} = \frac{2.0294}{1.9706} = \mathbf{1.0298}.$$

Note that the signs change when a is negative.

EXAMPLE 7   Find c(2.3681).

We can use π as a base and since 3.1416 – 2.3681 ≈ 0.77, which is near to the 0.9273 triple we have c(π– 0.9273) = c(2.2143) = ½.

∴ c' = ½ and a = 0.1538.

$$\therefore \ c(2.3681) = \frac{2 \times \frac{1}{2} - 0.1538}{2 + \frac{1}{2} \times 0.1538} = \frac{0.8462}{2.0769} = \mathbf{0.4073}.$$

EXAMPLE 8   Find c(5.6968).

We know c(5.6968) = c(5.6968 – 2π) = c(–0.5864) = –c(0.5864) see Page 45.
So we use c'=3 and a = –0.0571.

$$\therefore \ c(5.6968) = -\frac{2 \times 3 + 0.0571}{2 - 3 \times 0.0571} = -\frac{6.0571}{1.8287} = \mathbf{-3.3122}.$$

EXAMPLE 9   Find A(5.691).

We have to find the angle in the triple whose code number is 5.691.
c'=6 is the closest known perfect triple, so A'=0.3303.

$$\therefore \ A(5.691) = A' + \frac{2(c'-c)}{c'c+1} = 0.3303 + \frac{2(6-5.691)}{6 \times 5.691+1} = 0.3303 + \frac{0.618}{35.146} = \mathbf{0.3479}.$$

EXAMPLE 10   Find A(2.615).

$$c'=3 \quad \therefore \ A(2.615) = 0.6435 + \frac{2(3-2.615)}{3 \times 2.615+1} = 0.6435 + \frac{0.7700}{8.845} = \mathbf{0.7306}.$$

EXAMPLE 11   Find A(1.016).

$$c'=1 \quad \therefore \ A(1.016) = 1.5708 + \frac{2(1-1.016)}{1 \times 1.016+1} = \mathbf{1.5549}.$$

EXAMPLE 12 Find A(0.362).

0.362 is close to $\frac{1}{3}$ and $c(\frac{1}{3}) = \pi - 0.6435$, so $c' = \frac{1}{3}$.

$$\therefore \; A(0.362) = 3.1416 - 0.6435 + \frac{2(\frac{1}{3} - 0.362)}{\frac{1}{3} \times 0.362 + 1} = \mathbf{2.4469}.$$

EXAMPLE 13 Find A(−3.201).

We will find A(3.201) and subtract the result from 2π.

$c' = 3 \qquad \therefore \; A(-3.201) = 2\pi - \{0.6435 + \frac{2(3-3.201)}{3 \times 3.201 + 1}\} = \mathbf{5.6776}.$

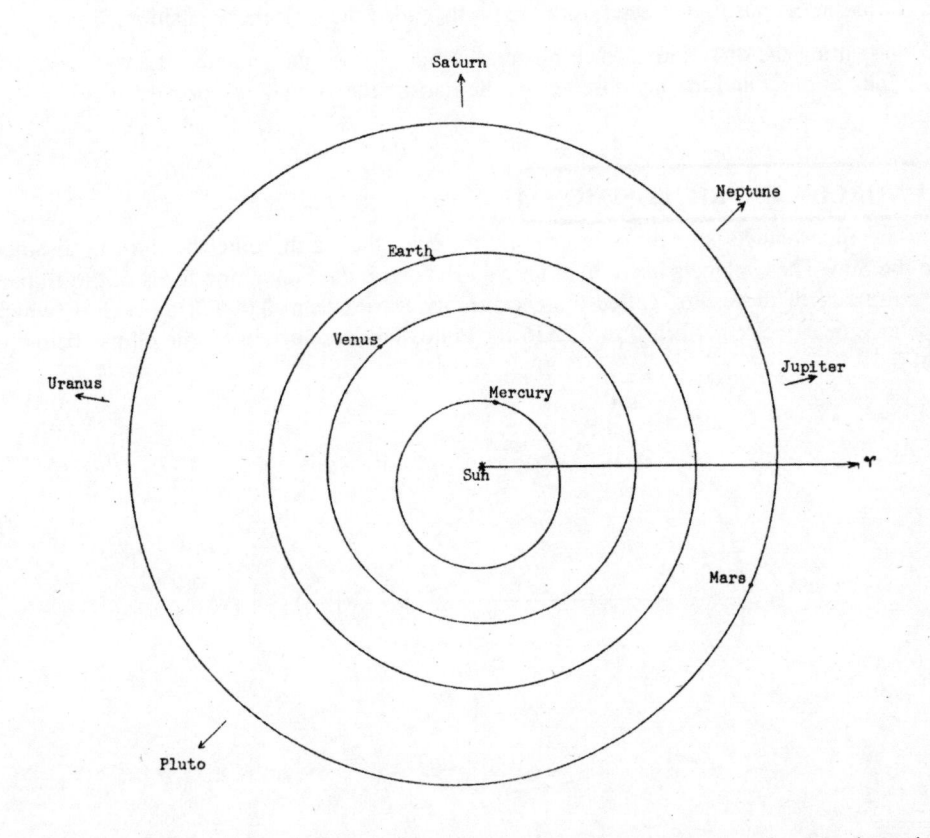

The orbits of Mercury, Venus Earth and Mars are shown here represented approximately by circles. ♈ is the zero or reference direction and the point marked on each orbit is its perihelion (the point of the orbit closest to the Sun) or the direction of its perihelion for the other planets.

On this scale Jupiter, Saturn, Uranus and Neptune would be 20, 35, 70, 110 cm from the Sun. The Sun's diameter would be less than ½ mm and the nearest star would be about 10 km away.

Chapter 5

PREDICTION OF PLANETARY POSITIONS

Here we consider how, given certain basic data, the position of a planet at some instant can be found. This is in three parts:

(1) Finding the position of a planet in its orbit around the Sun, called the heliocentric position.

(2) Finding its position as seen from the Earth, called the geocentric position.

(3) Locating the direction of each planet, the Sun, Moon and galaxies at any time of the day or night and for any observer on the Earth, relative to their horizon.

5.1   HELIOCENTRIC POSITION

There are nine planets known to be orbiting the Sun, the Earth being the third in distance from the Sun. These planets move in elliptical orbits with the Sun at one focus of the ellipse. The ellipticity of the orbits, called the eccentricity, varies from 0.00678 for Venus (which indicates a nearly circular orbit) to 0.2536 for Pluto. The eccentricity of the ellipse below is about 0.65.

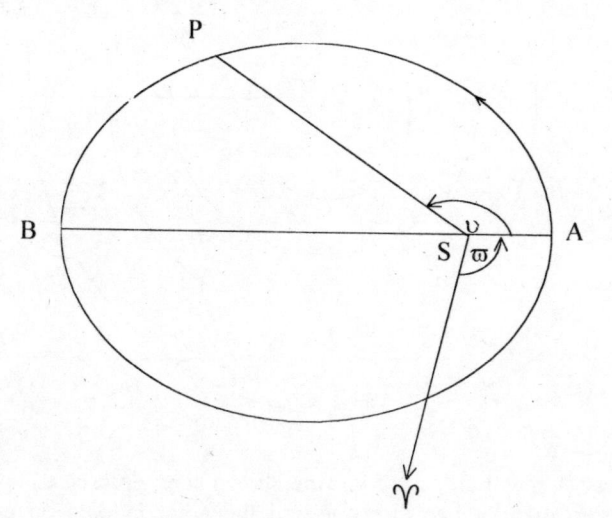

Figure 1: An elliptical orbit and the relationship of a planet, P, to the perihelion, A, and to the reference point ♈.

In Figure 1, S denotes the position of the Sun and P the position of a planet at some instant.

The perihelion point, A, is the point in the orbit which is closest to the Sun. Then $\stackrel{\wedge}{A}SP$ describes the planet's position in its orbit at that time relative to A and is called the true anomaly, v, of the planet.

However, since the various planetary orbits have their perihelia in different directions (see diagram on Page 53), a common point of reference is needed from which to measure the planetary positions. This reference point is called the first point of Aries, or the vernal equinox and is denoted by the symbol Υ (see Figure 1). It is defined in the next section and always lies in the ecliptic: the plane of the Earth's orbit. Thus given the longitude of perihelion, ϖ, for a planet, the heliocentric longitude, L, is given by $L = \varpi + v$. ϖ is almost constant for each planet and its value for each planet on 1985 January 0.5 are given in the table on Page 69. We shall assume for the moment that Υ is in the plane of the planetary orbits (see Example 4 and Appendix I).

If we use the aphelion point, B, as a base instead of A then ϖ and v will be measured as shown in Figure 2, Example 5.

ϖ being known, the problem is really to find v. The basic data from which we calculate v for a particular planet at a particular time are:

e — the eccentricity of the ellipse,
T — the time, called the periodic time, or period, taken by the planet to complete a circuit around the Sun, relative to the stars,
τ — the time at which the planet last passed through perihelion or aphelion,

(the values of e, T and some values of τ are given in the table on Page 69)

t — the time at which we require the planet's position.

The Mean Anomaly

We now define a quantity called the mean anomaly, M (already discussed on Pages 31,2) for a planet with period T, at time t. This is the angle, measured in radians, through which it would have moved since passing perihelion if it were moving at a constant speed in a circular orbit, with period T.

Thus, $\dfrac{M}{2\pi} = \dfrac{t-\tau}{T}$, $t - \tau$ being the time that has elapsed since perihelion passage.

$\therefore M = \frac{2\pi}{T}(t - \tau)$.

If, for example, we require M for a planet with a period of 300 days and 30 days have elapsed since perihelion passage, then $M = \frac{2\pi}{300} \times 30 = 0.6283$ radians.

Once M has been calculated its code number c(M) is found. Then c(v), the code number of v is given approximately by:

$c(v) = (1 - 2e)c(M)$ derived in Appendix II.

We then decode $c(v)$ to get v.

EXAMPLE 1   Find the heliocentric longitude of the Earth on 1985 Feb 14 0^h.

From Page 69 our basic data is:
  e = 0.01672
  T = 365.26
  ϖ = 103.413°
  τ = 1985 Jan 3 20^h
  t = 1985 Feb 14 0^h

$\therefore$ $M = \frac{2\pi}{T}(t - \tau) = \frac{2\pi}{365.26}(41.17) = 0.7082$ radians, as $41\frac{1}{6}$ days have elapsed since perihelion passage.

Then $c(0.7082) = 2.7051$ see Page 51, Example 5.

$\therefore$ using $c(v) = (1 - 2e)c(M)$,
        $c(v) = 0.96656 \times 2.7051 = 2.615$.

Then $A(2.615) = 0.7306$ see Page 52, Example 10.

$\therefore$ $v = 0.7306$ radians = 41.86°.

$\therefore$ $L = \varpi + v = 103.413 + 41.86 = 145.27°$ or $145°16'$.

Note: (1) This process can be summarised as follows:

      (i)   find M using $M = \frac{2\pi}{T}(t - \tau)$,
      (ii) find $c(M)$,
      (iii) find $c(v)$ using $c(v) = (1 - 2e)c(M)$,
      (iv) decode to get v,
      (v) find L using $L = \varpi + v$.

(2) The formula $c(v) = (1 - 2e)c(M)$ is approximate and in the derivation in Appendix II more terms are obtained so that if the formula for $c(v)$ is not accurate enough a more accurate one can be used. For example, if the e^2 term is included:

$$c(v) = \{1 - 2e + e^2(1.5 + \frac{1}{c^2 + 1})\}c \qquad \text{where } c = c(M)$$

In fact the answer obtained in Example 1 is 1 arc minute from the true value of 145°15' (to the nearest minute) and the formula above would give this value. See also Page 59 Note, in which a further simplification is indicated.

(3) The formulae normally used for calculating the true anomaly are:

$$M = E - e\sin E \text{ and } \tan\frac{v}{2} = \sqrt{\frac{1+e}{1-e}} \tan\frac{E}{2}$$

so that the eccentric anomaly, E, must be obtained from the first formula and substituted into the second. But the first equation, known as Kepler's equation and discussed in Chapter 3, is transcendental and so cannot be solved directly. This means that a great deal of work is involved in finding v, and an electronic calculating device is normally essential. However the simple formula given here, though it is not exact like the two formulae quoted above, gives us the longitude without the need for a calculating device.

EXAMPLE 2   Find the heliocentric longitude of Saturn on 1985 Feb 14 0^h.

Our data is: e = 0.05560
T = 10759
ϖ = 91.824°
τ = 1974 Jan 8 0^h
t = 1985 Feb 14 0^h

$\therefore M = \frac{2\pi}{10759}(4055) = 2.3681$ radians.

Then c(2.3681) = 0.4073 see Page 52, Example 7.

$\therefore c(v) = 0.8888 \times 0.4073 = 0.3620.$

And A(0.362) = 2.4469 see Page 53, Example 12.

$\therefore v = 140.20°.$

$\therefore L = 91.824 + 140.20 = 232.02°$ or 232°1'.

EXAMPLE 3   Find the heliocentric longitude of the Earth on 1985 Dec 1 0^h.

We have: e = 0.01672
T = 365.26
ϖ = 103.413°
τ = 1985 Jan 3 20^h
t = 1985 Dec 1 0^h

$\therefore M = \frac{2\pi}{365.26}(331.17) = 5.6968$ radians.

But $c(5.6968) = -3.3122$ see Page 52, Example 8
$\therefore$ $c(v) = 0.96656 \times -3.3122 = -3.2014$.
Then $A(-3.201) = 5.6776$ see Page 53, Example 13.

$\therefore$ $v = 325.30°$.

$\therefore$ $L = 103.413 + 325.30 = 68.72°$ or $68°43'$.

Here the Earth is nearly back at its perihelion point and we could also find L by using the 1986 perihelion passage as a base.

As in Example 1 (and for the same reason) this longitude differs by 1 arc minute from the true value of $68°44'$. Example 2 is correct. Note that in this chapter by 'true' or 'correct' value we mean correct to the nearest minute of arc.

EXAMPLE 4   Find the heliocentric longitude of Venus on 1985 Dec 1 0^h.

We have: $e = 0.00678$
$\qquad\qquad T = 224.70$
$\qquad\qquad \varpi = 131.304°$
$\qquad\qquad \tau = 1985$ Oct 6 21^h
$\qquad\qquad t = 1985$ Dec 1 0^h

So $M = \frac{2\pi}{224.7}\,(55.125) = 1.5414$ radians.

   $c(1.5414) = 1.0298$ see Page 52, Example 6.

$\therefore c(v) = 0.9864 \times 1.0298 = 1.0158$.

   $A(1.016) = 1.5549$ see Page 52, Example 11.

$\therefore v = 89.09°$.

$\therefore$ $L = 131.304 + 89.09 = 220.394°$ or $220°24'$.

Because of the inclination of the orbit of Venus to the plane of the ecliptic this answer is actually composed of two angles which are not in the same plane. This is explained in Appendix I, and the longitude measured along the ecliptic (i.e. in one plane) is calculated. This is $220°27'$.

EXAMPLE 5   Find the heliocentric longitude of Mars on 1985 Dec 1 0^h.

The data is: $e = 0.09339$
$T = 687.0$
$\varpi = 155.802°$
$\tau = 1985$ Oct 17 1^h
$t = 1985$ Dec 1 0^h

We see that Mars has recently passed its aphelion point and we can use this as our base:

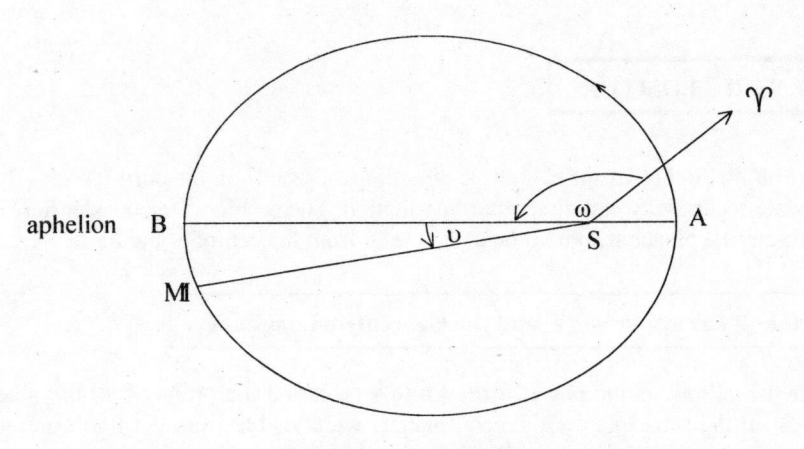

Figure 2: Using aphelion as base.

The procedure is the same as before except that the formula connecting the code numbers of v, M becomes: $c(v) = (1 + 2e)c(M)$ derived in Appendix II:

Then $M = \frac{2\pi}{687}(44.96) = 0.4112$ radians.

  $c(0.4112) = 4.7952$ see Page 51, Example 4.

$\therefore$ $c(v) = 1.1868 \times 4.7952 = 5.6909$

  $A(5.691) = 0.3479$ see Page 52, Example 9.

$\therefore$ $v = 19.93°$.

$\therefore$ $L = 155.802 + 19.93 = 175.73°$ or $175°43'$.

The correct value is $175°31'$ so that this answer is 12' too large. This is because of the relatively large eccentricity of the orbit of Mars. However we can calculate $c(v)$ more accurately using the formula: $c(v) = \{1 - 2e + e^2(1.5 + \frac{1}{c^2+1})\}c$. See Page 56, Note 2 and Appendix II.

With this formula we get $c(v) = 5.755$ and the true value of L is obtained.

Note: Since c(M) can always be numerically greater than 1 this formula can be written as
c(v) = {1 + 2e + 1.5e^2}c(M) using aphelion as base, or
c(v) = {1 − 2e + 1.5e^2}c(M) using perihelion as base.

Thus we can say that c(v) is proportional to c(M) for each planet. These constants of proportionality are listed in Page 69.

In the last example we would then have c(v) = 1.2 × 4.795, which leads to the correct value, L = 175°31'.

5.2   GEOCENTRIC POSITION

The planets all orbit the Sun in an anticlockwise direction (as seen from the north), each orbit having its own size, eccentricity and direction of perihelion. The problem we consider here is locating the position of a planet at some time as it is seen from the centre of the Earth.

Definition of the Reference Point ♈ and the Geocentric Longitude

As stated earlier the ecliptic is the plane of the Earth's orbit and the orbits of all the other planets are almost in the same plane: if several planets were visible from the Earth on one night an observer would see that they form a line across the sky: this line is the ecliptic.

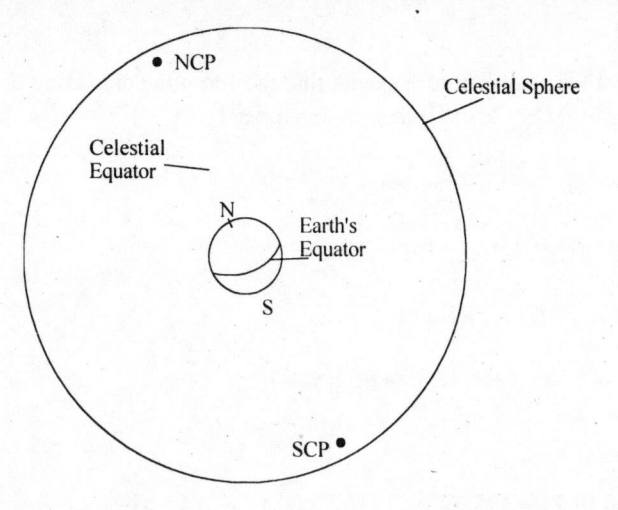

Figure 3: The Celestial Equator and the North and South Celestial Poles, NCP and SCP.

The Earth spins once a day on its axis, so that it has north and south poles (N, S in Figure 3) and an equator. If these poles and equator were projected outwards from the centre of the Earth onto the celestial sphere they would give the north and south celestial poles and the celestial equator (the north celestial pole being close to the pole star). These may be considered to be fixed in position on the celestial sphere (there is a slow movement due to precession, but we will not consider this here).

We now have two circles on the celestial sphere: the ecliptic and the celestial equator:

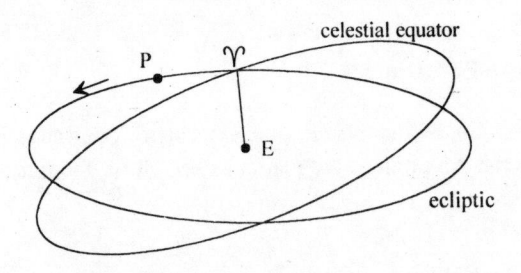

Figure 4: The Ecliptic and the Celestial Equator defining ♈.
The Earth is at the centre.

Disregarding the rotation of the Earth, the Sun appears to move along the ecliptic, completing a circuit in one year. The ecliptic and the celestial equator intersect in two points, and the one through which the Sun passes in going from south to north of the equator is called the first point of Aries, denoted by ♈. This is the zero or reference point for all planetary positions. All longitudes, heliocentric and geocentric are measured along the ecliptic from ♈. Thus if a planet has a geocentric longitude of 30° this means that the planet will be found 30° from ♈ measured eastwards (i.e. leftwards when facing ♈) along the ecliptic. This would be angle ♈EP in Figure 4.

We find the geocentric longitude of a planet by first finding its heliocentric longitude and adding an angle called the geocentric correction.

Finding the Geocentric Correction θ

Consider the orbits of the Earth and an inner planet (Figure 5a) and of the Earth and an outer planet (Figure 5b). Though we represent the orbits as concentric circles here, this is of course not the case.

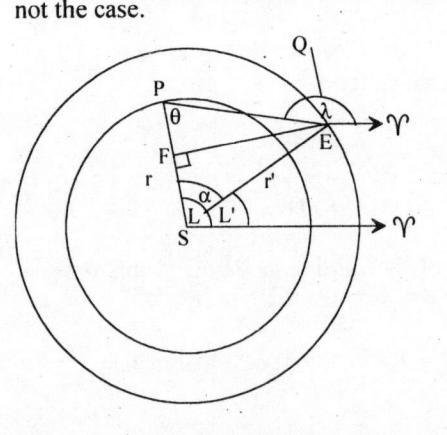

Figure 5a: Inner planet and Earth's Orbit.

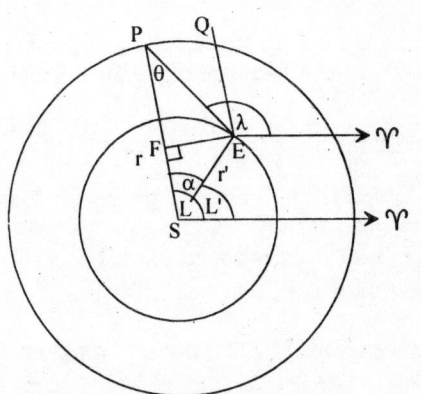

Figure 5b: Outer planet and Earth's Orbit .

The following derivation holds for both Figure 5a and Figure 5b.

Let L, L' be the heliocentric longitudes of the planet P and the Earth E respectively at some instant, and let r, r' be their distances from the Sun..

Also let $L - L' = \alpha$, $E\hat{P}S = \theta$ and $\gamma\hat{E}P = \lambda$.

The geocentric position of the planet (its geocentric longitude) is the angle between the direction of γ and the direction of the planet as seen form the Earth, i.e. it is the angle λ.

Draw EQ parallel to SP.

Then $P\hat{E}Q = \theta$ and $\gamma\hat{E}Q = L$.

Therefore, $\lambda = L + \theta$.

Thus θ is the correction which must be made to the heliocentric longitude, L, of the planet to get its geocentric longitude, λ.

Now drop a perpendicular EF from E onto SP. Let x,y,z be a triple representing α, and let x,y,z = xn,yn,zn = xn,yn,r' where r' = zn.

Then EF = yn and SF = xn.

Therefore, $c(2\theta) = \frac{r-xn}{yn} = \frac{\frac{zr}{r'} - \frac{zxn}{r'}}{\frac{zyn}{r'}} = $ since $c(2\theta) = \cot\theta$ (see Page 47, Note).

And r' = zn, so $c(2\theta) = \frac{zR-x}{y}$, where $R = \frac{r}{r'}$.

This means that we can find θ by first subtracting the heliocentric longitude of the Earth from the heliocentric longitude of the planet, to get α; finding a triple x, y, z representing α and then finding R, the ratio of the distances of the planet and the Earth from the Sun. This gives $c(2\theta)$, from which we find θ by the method previously used (see Pages 51-53).

We shall use the mean values of r, r' here (given as 'a' on Page 69), and in Appendix III we show how to calculate their values accurately for any given time.

EXAMPLE 6   Find the geocentric longitude of Venus on 1985 Dec 1 0^h.

We have already found the heliocentric longitudes of the Earth and Venus at this time in Examples 3, 4.

So we have L = 220°27' and L' = 68°44', so that $\alpha = L - L' = 151°43' = 2.6480$ radians.

We now find a suitable triple for this angle:

$2.6480 = \pi - 0.4936$, $\therefore$ $c(2.6480) \approx 1,4$ (cf Example 7, Page 52)

Thus, the triple $- 15,8,17$ can represent the angle α, so that $x = -15$, $y = 8$, $z = 17$.

Also, from Page 69, $a = 0.7233$, $a' = 1.0000$ so that $R = 0.7233$.

$\therefore$ $c(2\theta) = \frac{0.7233 \times 17 - -15}{8} = 3.412$.

$\therefore$ $A(3.412) = 0.6435 + \frac{2(3-3.412)}{3 \times 3.412 + 1} = 0.5702$ radians. (cf Example 9, Page 52)

$\therefore$ $2\theta = 0.5702$ radians and so $\theta = 16.334°$.

$\therefore$ $\lambda = 220°27' + 16°20'$.

$\therefore$ $\lambda = \mathbf{236°47'}$.

The correct result is $236°48'$.

EXAMPLE 7 Find the geocentric longitude of Saturn on 1985 Feb 14 0^h.

From Examples 2, 1 we have $L = 232°1'$ and $L' = 145°15'$,
so that $\alpha = 86°46' = 1.5144$ radians.

Then $c(1.5144) = c(\frac{\pi}{2} - 0.0564)$ and $\frac{2}{0.0564} \approx 35$.

So $c(1.5144) = 1 - 35 = \frac{36}{34} = 18,17$.

Then $x = 18^2 - 17^2 = 35$, $y = 2 \times 18 \times 17 = 612$, $z = 18^2 + 17^2 = 613$.

(Or, more directly, having got the code number 35 above we can write down the triple with this code number (see Page 45) and interchange the first two elements to get the complementary triple).

Also, from Page 69, $R = 9.539$.

$\therefore$ $c(2\theta) = \frac{9.539 \times 613 - 35}{612} = 9.497$.

$\therefore$ $A(9.497) = 0.2213 + \frac{2(9-9.497)}{9 \times 9.497 + 1} = 0.2098$ radians.

$\therefore$ $\theta = 6°1'$.
$\therefore$ $\lambda = 232°1' + 6°1' = \mathbf{238°2'}$.

The correct value here is $237°44'$. In Appendix II we show how to calculate R to greater accuracy so that the correct value can be obtained.

EXAMPLE 8   Find the geocentric longitude of Mars on 1985 Dec 1 0^h.

From Examples 5, 3 we have L = 175°31' and L' = 68°44',
so that α = 106°47' = 1.8637 radians.

Then $c(1.8637) = c(\frac{\pi}{2} + 0.2929) \approx 1 + 7 = 3.4$.

Then x = –7, y = 24, z = 25.

And since R = 1.5237

$\therefore c(2\theta) = \frac{1.5237 \times 25 - -7}{24} = 1.879$.

$\therefore A(1.879) = 0.9273 + \frac{2(2-1.879)}{2\times1.879+1} = 0.9782$ radians.

$\therefore \theta = 28°1'$.

$\therefore \lambda = 175°31' + 28°1' = \mathbf{203°32'}$.

We show in Appendix III how to obtain the correct value 201°27' by using better values for R and $c(\alpha)$.

EXAMPLE 9   Find the geocentric longitude of the Sun on 1985 Dec 1 0^h.

We do not need to·make any geocentric correction in this case as
the geocentric longitude of the Sun = heliocentric longitude of the Earth + 180°.

That is: λ(Sun) = L(Earth) + 180°.

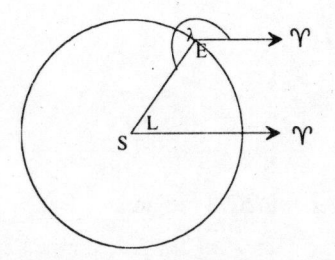

The geocentric longitude of the Earth on 1985 Dec 1 0^h was found in Example 3 to be 68°44'.

$\therefore \quad \lambda = 68°44' + 180° = \mathbf{248°44'}$.

EXAMPLE 10 Find the geocentric longitude of the Moon on 1985 Dec 16 0^h.

The Moon's behaviour is notoriously erratic: Newton complained that the problem of predicting the Moon's behaviour made his head ache and kept him awake at night. This is mainly because the Moon is influenced by two bodies: the Earth and the Sun, whereas the planets are influenced mainly by the Sun and very little by other planets. Also the Moon's proximity to the earth means that slight variations in its motion are very noticeable. The two main effects which disturb the Moon from its otherwise elliptical orbit around the Earth are:

(i) the evection, due to the gravitational pull of the Sun, and
(ii) the variation, due to the variation of the Sun's pull during a lunar month.

So we cannot expect to achieve any great accuracy when finding the Moon's longitude with our simple formula. However, the Moon does orbit the Earth so that we have only one calculation to make: using $c(v) = (1 - 2e)c(M)$.

Our basic data is: $e = 0.0549$
$T = 27.322$
$\varpi = 244.69°$
$\tau = 1985$ Dec 11 $0^h\ 32^m$
$t = 1985$ Dec 16 0^h

Then $M = \frac{2\pi}{27.322} \times 4.978 = 1.1447$,

$c(1.1447) = 1.552$,

and $c(v) = (1 - 0.1098) \times 1.552 = 1.382$.

$\therefore\ v = 1.253$ radians $= 71.79°$.

Therefore, $\lambda = 71.79 + 244.69 = \textbf{316°29'}$.

The correct value is $317°15'$.

5.3   THE PLANET FINDER

Geocentric longitudes, the longitudes calculated in the previous chapter, are the quantities found listed in an astronomical Ephemeris. But they do not tell a potential observer somewhere on the Earth where the planets are at some particular time. The Planet Finder is a simple cardboard device that will give the approximate direction in space of the planets, Sun, Moon and the Galactic Centre.

We would like to be able to ascertain the direction of any planet at any time, for any observer on the Earth. The accurate solution to this problem is found using geocentric longitudes and spherical geometry. However, if we are content with an approximate answer, we can quickly and easily locate the direction of the planets and also the plane and centre of our Galaxy. We assume that we begin with the geocentric longitudes of the planets which we wish to locate: found either by calculation, as has been shown, or from an Ephemeris. We also need to know either the observer's north direction and the approximate latitude of the observer, or, (in the northern hemisphere) the direction of the Pole Star.

We shall first explain the construction of the Planet Finder, then how to use it, but we may mention here that once the orientation of the two main circles and their motion is understood, it is possible to dispense with the model and to imagine their position on the sky. This does not require a lot of practice.

The Finder consists of three intersecting circles which are inclined to each other at different angles. The three circles represent:

(1) The plane of the equator.

(2) The ecliptic: the plane of the Earth's orbit.

(3) The plane of the Galaxy.

Construction

From thin cardboard cut out three circles about the size of those in Figures 1, 2 and 3 on Pages 131-135 and copy the various angles and signs, etc. shown, or detach the pages from the book and stick them onto the card. Cut out the inside of circles 2 and 3 so that two rings about 2½ cm wide are obtained. Then cut four narrow slots in each of the three circles along the dashed lines at A, B, C, D, E, F. The widths of the slots should be about equal to the thickness of the card used.

Take circles 1 and 2 and slot them together so that the slots marked A are together, those marked B are together and the signs from Aries to Virgo are above circle 1. Then slot circle 3 onto circles 1 and 2 so that the slots C, D, E, F on circle 3 meet slots C, D, E, F on circles 1 and 2. Circles 1 and 2 should then be inclined at about 23° to each other, and circles 1 and 3 should be inclined at about 63° to each other.

The Finder is now ready for use.

Application

First the planets whose direction we require should be shown in their correct positions on the ecliptic circle. The positions may be obtained by calculation or by use of an Ephemeris.

Suppose we want to know where all the planets, including the Sun and Moon, are on 1985 Dec 1 0^h for an observer in London. The longitudes are:

SUN	249°	JUPITER	312°
MOON	107°	SATURN	242°
MERCURY	244°	URANUS	258°
VENUS	237°	NEPTUNE	272°
MARS	201°	PLUTO	216°

One way of indicating these positions is to push a pin through the ecliptic circle at the appropriate place, with the planet's symbol or name attached.

When the planets are all in position push a drawing pin or nail through the centre of the Finder from the top, so that by holding on to it with one hand the Finder can be rotated with the other. Alternatively, instead of holding the pin or nail, it can be pushed into the end of a stick: it will then be easier to hold and rotate the Finder.

Next, if the position of the Pole Star is known, point the Finder at it: i.e. the axis of the central pin should point towards the Pole Star. Note that the direction of the Pole Star from any position on the Earth does not change, day or night (except over very long periods of time). If the position of the Pole Star is not known, hold the Finder at arm's length pointing due north (so that the axis of the central pin is horizontal), and then raise the arm through an angle equal to the observer's latitude (the latitude of London is about 52°). The Finder should then be pointing to the Pole Star.

Now the direction of the Sun is known because for our example the time is midnight and so the Sun will be at its lowest point. With the Finder still pointing at the Pole Star we therefore rotate it so that the Sun marker is at its lowest point. Then the direction of each of the other planets is indicated: a line from the centre of the Finder through the Moon's position on the Finder and extended indefinitely outwards gives the direction of the Moon, and so on. The plane of the Galaxy and the direction of the Galactic Centre are also shown.

In this example it would be found that all the planets cluster around the Sun and are therefore not visible at this time, except for the Moon which according to the model should high in the sky towards the south-east. Note that any planet below the level of the centre of the Finder is below the horizon, and any planet above it is above the horizon.

Four hours later, at 4 a.m. the positions of the planets are found by rotating the Finder by 60° in a clockwise direction from above, as the earth rotates 15° per hour. Or since circle 1 is marked in hours, rotate by 4 hours. It should then be found that Mars is rising in the east, and that the Moon is now at its highest point and due south.

Similarly, if the planetary positions were required at 11 a.m. on the same day rotate the Finder so that the Sun is at its highest point (it should then be due south) and rotate 1 hour backwards, i.e. anticlockwise. Jupiter should be just above the horizon.

Thus, there are four steps in using the Planet Finder:

(1)   Align the planets in their correct positions on the ecliptic circle.

(2)   Point the Finder at the Pole Star.

(3)   Rotate so that the Sun is at its highest or lowest point, depending on whether the time is nearest to 12 noon or 12 midnight.

(4)   Rotate the Finder clockwise for times after 12 o'clock and anti-clockwise for times before 12 o'clock (15° for each hour).

The same procedure could also be used for an observer in the southern hemisphere, except that the pin pushed through the centre of the Finder should be pushed from the other side. The user should then point the Finder towards the south celestial pole.

Note that during Summer Time, if an hour has been added (as it is in the case of British Summer Time) an hour must first be subtracted from clock time, and then the difference, as before, between this and 12 o'clock found.

Though the methods shown in this chapter show that it is possible to calculate positions they are probably of little practical use as all the information is readily available in Ephemerides and Almanacs. But they do show that it is possible to predict positions of heavenly bodies without too much effort and without a calculator. If a lesser degree of accuracy was permissible calculations could of course be further reduced.

TABLE OF PLANETARY DATA

	Time of τ		Longitude of Per. or Aph. ϖ		T Sidereal Period (days)	a Mean dist. from Primary (au)	e Eccentricity	$1-2e+\frac{3}{2}e^2$ (Per.)	$1+2e+\frac{3}{2}e^2$ (Aph.)
	Perihelion	Aphelion	true	mean					
MERCURY	1985 Sep 3 14^h19^m 1985 Nov 30 13^h34^m	1985 Oct 17 13^h56^m	77°.238 257°.237 77°.241	77°.222 257°.222 77°.222	87.97	0.3871	0.2056	0.6522	1.4746
VENUS	1985 Feb 24 6^h 1985 Oct 6 21^h	1985 Jun 12 12^h24^m	131.419 311.285 131.304	131.360 311.360 131.360	224.70	0.7233	0.00678	0.9865	1.0136
EARTH	1985 Jan 3 20^h	1985 Jul 5 10^h	103.413 283.328	102.682 282.682	365.26	1.0000	0.01672	0.9670	1.0339
MARS	1984 Nov 7 10^h15^m	1985 Oct 17 1^h	335.724 155.802	335.783 155.783	687.0	1.5237	0.09339	0.8263	1.2000
JUPITER	1975 Aug 12 14^h	1981 Jul 28 7^h	13.962 194.873	14.081 194.081	4333	5.2028	0.04848	0.9066	1.1005
SATURN	1974 Jan 8 0^h		91.824	92.754	10759	9.539	0.05560	0.8934	1.1158
URANUS	1966 May 20 0^h		168.528	170.418	30685	19.182	0.04728	0.9088	1.0979
NEPTUNE		1959 Jan 26 0^h	215.3	224.495	60191	30.06	0.00859	0.9829	1.0173
PLUTO	1741 Oct 24 0^h		223.7	224.250	90465	39.71	0.2536	0.5893	1.6037
MOON	1985 Dec 11 0^h32^m		244.69		27.322	0.002570	0.0549	0.8947	1.1143

au = astronomical units = in units of the Earth's mean distance from the Sun

Values of τ subsequent to those listed in this Table can be found approximately by adding T, the sidereal period, or multiples of T. In that case the mean values of ϖ should be used.

SPHERICAL TRIANGLES

Spherical trigonometry has many applications in navigation, aviation and geography etc. as well as in astronomy.

6.1   SPHERICAL TRIANGLES USING TRIPLES

A spherical triangle ABC is a triangle on the surface of a sphere, the sides being arcs of great circles (a great circle is a circle on the surface of a sphere whose centre is also the centre of the sphere). It has three sides and three angles, the sides being expressed by angles:

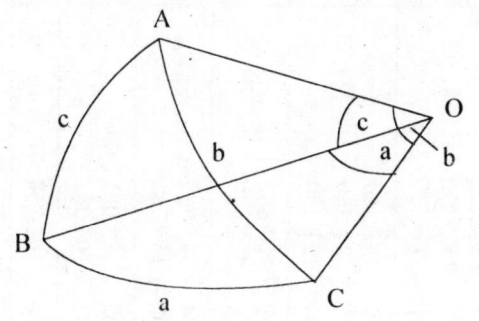

O is the centre of the sphere, and $a = B\hat{O}C$, $b = A\hat{O}C$, $c = A\hat{O}B$.

By $\hat{A}$ we mean the angle between the tangents to the arcs AB and AC at A, or, which amounts to the same thing, the angle between the planes AOB and AOC. Similarly for $\hat{B}$ and $\hat{C}$.

Triple Notation for Spherical Triangles

These six angles a, b, c, A, B, C can each be described by a triple and we can use the following arrangement to show them:

```
a |  •   •   •        A |  •   •   •
b |  •   •   •        B |  •   •   •
c |  •   •   •        C |  •   •   •
```

Three triples suffice to define a particular spherical triangle.

For example:

a	4	3	5		A	♦	♦	♦
b	5	12	13		B	♦	♦	♦
c	♦	♦	♦		C	8	15	17

and the solution of the triangle consists in finding the missing numbers in the chart. We will normally consider that it will be sufficient to find just two elements of a triple (unless the third is easily found or is needed in further work).

Standard Formulae of Spherical Trigonometry

We need to know the relations that exist between the 18 elements in the chart. These relations are normally expressed by the following four formulae:

The Cosine Rule $\quad \cos a = \cos b \cos c + \sin b \sin c \cos A$

The Sine Rule $\quad \dfrac{\sin a}{\sin A} = \dfrac{\sin b}{\sin B}$

The Cotangent (or Four Parts) Rule $\quad \cos a \cos C = \sin a \cot b - \cot B \sin C$

The Polar Cosine Rule $\quad \cos A = \sin B \sin C \cos a - \cos B \cos c$

These equations are somewhat complex and are not easy to remember or apply. However, when translated into triple form simple patterns emerge which enable us to solve spherical triangles much more easily.

The following additional points may be noted:

1. We will call a, b, c the sides, and A, B, C the angles of a spherical triangle.

2. The proofs of the formulae are given (except for the polar cosine rule) on Page 124-126.

3. Right-angled spherical triangles are dealt with later.

4. The following limitations on a, b, c, A, B, C may be of interest:
   (i) $a + b + c < 360°$,
   (ii) $180° < A + B + C < 540°$
   (iii) $a + b > c, \ a + c > b, \ b + c > a$.

The Cosine Rule to find an Angle

We take up first the problem of finding an angle when the three sides are given.

Suppose we have:

a	4	3	5		A	♦	♦	♦
b	12	5	13		B	x	y	r
c	15	8	17		C	♦	♦	♦

and we want B) x, y, r.

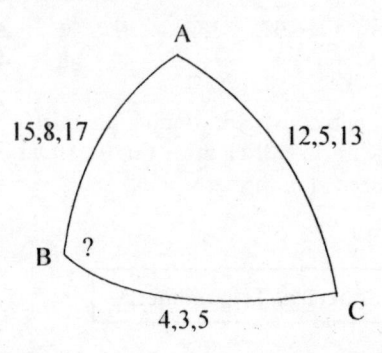

The ratio x:r (i.e. cos B : hence the name 'cosine rule') is given by the following pattern:

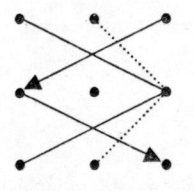

The lines in the diagram above indicate the numbers whose product is to be taken; the arrows indicate where we should start, because the two products '<' and '>' are to be subtracted. And the dashed lines indicate the product for the right-hand part of the ratio. Absence of the arrows in later patterns will indicate that the first two products are to be added.

So   x : r = 12.5.17 – 13.4.15 : 13.3.8 (where the dots indicate multiplication)
         = 10 : 13
    ∴ B) 10, – , 13.

The subtraction of the products 12.5.17 – 13.4.15 can be facilitated by mentally taking out any common factors i.e. 4.3.5(17 – 13), and any common factors which are also in the third product, 13.3.8, cancel out. In this case 3 and 4 cancel out, so we get
5(17 – 13) : 26 = 10 : 13.

For C : to maintain symmetry we write a, b, c in the order a, c, b (or b, c, a).

a	4	3	5			
c	15	8	17	C	x	y	r
b	12	5	13			

Then the same pattern of products gives 15.5.13 – 17.4.12 : 17.3.5
                    = 325 – 272 : 85 (cancelling the common factor, 3)
                    = 53 : 85

∴ C) 53, – , 85.

For A :
$$\begin{array}{r|rrr} b & 12 & 5 & 13 \\ a & 4 & 3 & 5 \\ c & 15 & 8 & 17 \end{array} \qquad \begin{array}{c|ccc} A & x & y & r \end{array}$$

$x : r = 4.13.17 - 5.12.15 : 5.5.8$
$\quad\quad = 221 - 225 : 50$
$\quad\quad = -4 : 50$

$\therefore$ A) $-2, -, 25$.

The minus here indicates the angle is obtuse.

If it ever happens that we get an answer in which $x > r$, so that x, y, r cannot be the elements of a triple, this is an indication that the given sides cannot form a triangle: see Page 71, 4 *(iii)*.

Note: If the vertices of a spherical triangle are given by three direction vectors we can solve the triangle by first finding the sides as shown in Ex 1 Page 115 and then finding the angles as shown above. But see also the method shown on Page 116 (Spherical Triangles).

The Cosine Rule to find a Side

We use this rule when we know two sides and the angle between them and we want to find the side opposite the angle. We use a symmetrical arrangement of the triples, as we did in the previous examples.

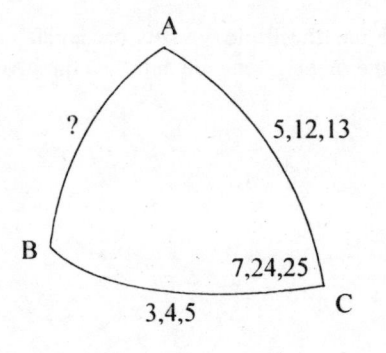

$$\begin{array}{r|rrr} a & 3 & 4 & 5 \\ C & 7 & 24 & 25 \\ b & 5 & 12 & 13 \end{array} \qquad \begin{array}{c|ccc} c & x & y & r \end{array}$$

The pattern is:

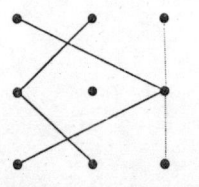

The two bold lines indicate products which are to be added this time; the dashed line again represents the right-hand portion of the ratio.

∴   x : r = 7.4.12 + 25.3.5 : 5.25.13
        = 711 : 1625

∴ A) 711, - , 1625.

The Sine Rule to find an Angle

Here we are given two sides and the angle opposite one of them and we want the angle opposite the other side.

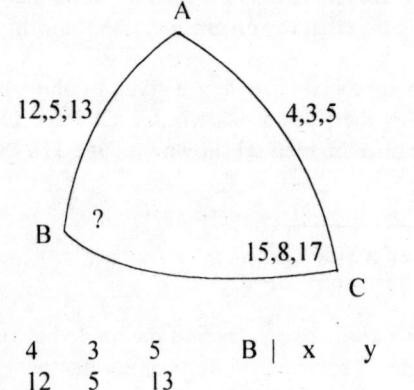

b	4	3	5	B	x	y	r
c	12	5	13				
C	15	8	17				

We know b, c, C and we use the triples in this particular order with the side opposite the required angle first, then the other given side and then the given angle.

The formula is:

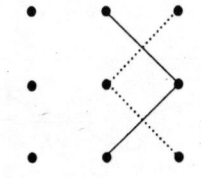

This formula gives the sine ratio of the angle.

Thus y : r = 13.3.8 : 5.5.17

∴ B) - , 312, 425.

The Sine Rule to find a Side

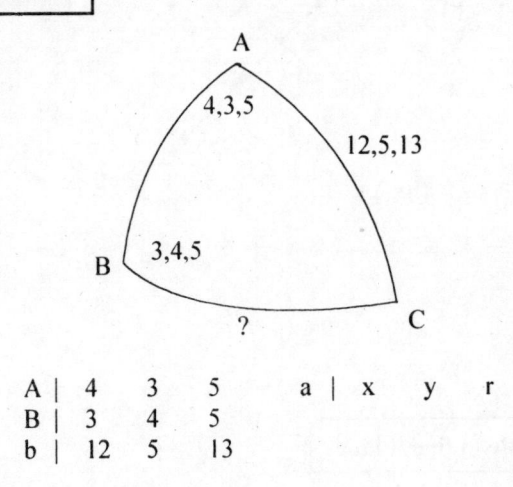

```
A | 4    3    5        a | x    y    r
B | 3    4    5
b | 12   5    13
```

We know A, B, b and we require a. We therefore put A first in the chart and then the other angle (keeping the two angles together just as we kept the two sides together in the last example).

With the Sine rule the pattern is the same whether finding a side or an angle.

So y : r = 5.3.5 : 4.5.13

∴ a) - , 15, 52.

The Cotangent Rule to find an Angle

We use this rule when we have two sides and the angle between them and we want another angle, or when we have two angles and the side between them and we want another side.

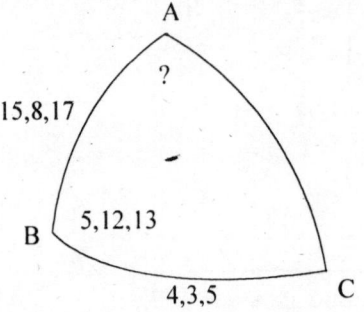

We know a, c, B and we order the letters as in the Sine Rule, with a first, as we require A, then the other side:

```
a | 4    3    5        A | x    y    r
c | 15   8    17
B | 5    12   13
```

The pattern is:

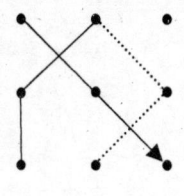

So x : y = 4.8.13 – 3.15.5 : 3.17.12
         = 191 : 612

∴ A) 191, 612, -

The Cotangent Rule to find a Side

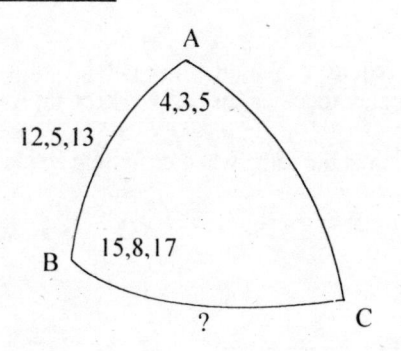

Also like the Sine Rule the pattern is the same as for finding an angle, except that we **add** the two products for instead of subtracting.

```
A |  4    3    5      a |  x    y    r
B |  15   8    17
c |  12   5    13
```

∴ x : y = 4.8.13 + 3.15.12 : 3.17.5

∴ a) 956, 255, -

The Polar Cosine Rule to find an Angle

If we have two angles and the common side and require the angle opposite the side, or if we have three angles and require a side, we use this rule.
The triples are arranged as for the Cosine Rule:

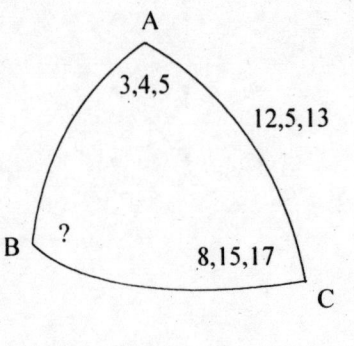

```
A |  3    4    5
b | 12    5   13      B | x    y    r
C |  8   15   17
```

The formula is:

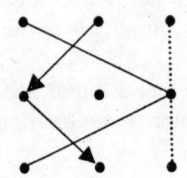

$\therefore$ x : r = 12.4.15 – 13.3.8 : 5.13.17
         = 408 : 1105

$\therefore$ B) 408, - , 1105

The Polar Cosine Rule to find a Side

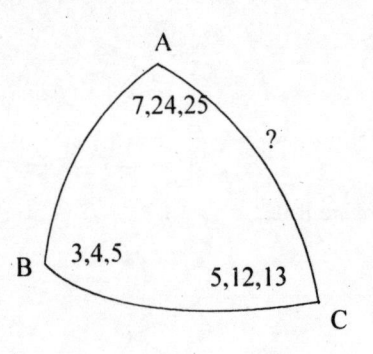

```
A | 7   24   25
B | 3    4    5       b | x    y    r
C | 5   12   13
```

Using:

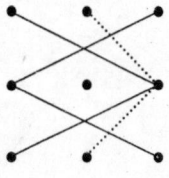

we get x : r = 3.25.13 + 5.7.5 : 5.24.12
            = 230 : 288

So b) 115, - , 144

It will be seen that these patterns are very similar to those for the Cosine Rule.

We may note two points relating to all the formulae.

(i) The starting point in each pattern is the first number on the same row as the required triple. (For the Sine Rule the first number is in the second column as the first column is not used).
(ii) It is only when finding an angle that a minus sign comes into the formula (except for the Sine Rule which has no addition or subtraction anyway).

Further Illustrations

EXAMPLE 1   Suppose we have the triangle below in which we know b, c, A and require a, B, C.

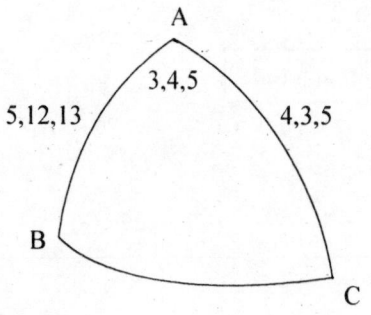

We can find a using the Cosine Rule:

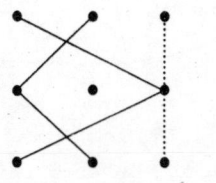

```
b | 4   3    5
A | 3   4    5        a |  208, - , 325 = 16, - , 25
c | 5   12   13
```

We can use the Cotangent Rule to find B and C:

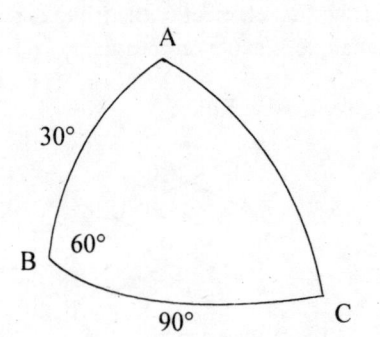

```
b | 4   3    5     B |  195, 156, - = 5, 4, -
c | 5   12   13
A | 3   4    5


c | 5   12   13    C |  –69, 240, - = –23, 80, -
b | 4   3    5
A | 3   4    5
```

We may check these results using the Cotangent Rule to find c:

```
C | –23   80      -       c |  –2300+6400, 240.41, - = 5, 12, 13.
B |  5     4     √41
a |  16   3√41    25
```

EXAMPLE 2 Solve:

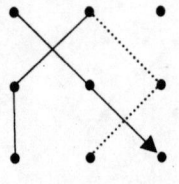

Recall:
```
30°  |  √3    1    2
45°  |  1     1    √2
60°  |  1     √3   2
90°  |  0     1    1
120° |  –1    √3   2
```

To find b we can use the Cosine Rule:

```
a |  0    1    1
B |  1   √3    2          b |  1, -, 4
c |  √3   1    2
```

We now use the Sine Rule to find C. We know c, b, B.

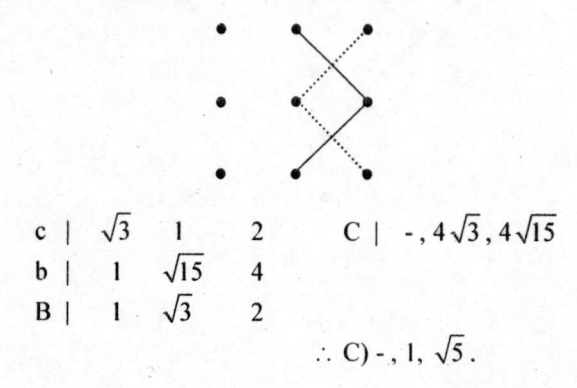

```
c |  √3    1     2          C |  -, 4√3, 4√15
b |  1    √15    4
B |  1    √3     2
                    ∴ C) -, 1, √5 .
```

However, in calculating the third element of the triple from the other two we do not know if it plus or minus, i.e. whether the angle is acute or obtuse. In this case we may use the Cotangent Rule:

```
c |  √3    1    2          C |  2√3, √3, - = 2, 1, √5
a |  0     1    1
B |  1    √3    2
```

In fact since $\sqrt{3}.1.2 > 1.0.1$ the first element of the triple is positive and this can be used as a test to determine whether an angle is acute or obtuse.

Angle A can be found using the Cosine Rule:

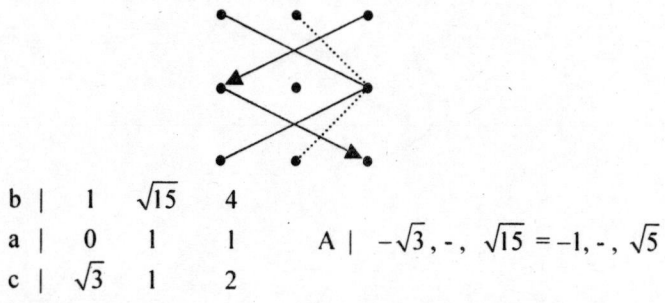

```
b |  1    √15    4
a |  0     1     1          A | -√3, -,  √15 = -1, -, √5
c |  √3    1     2
```

As a check we can find b from A, B, C using the Polar Cosine Rule:

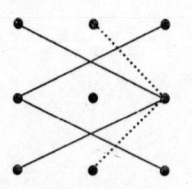

$$\begin{array}{c|ccc}
A & -1 & 2 & \sqrt{5} \\
B & 1 & \sqrt{3} & 2 \\
C & 2 & 1 & \sqrt{5}
\end{array} \qquad b \mid -4+5, -, 4 = 1, -, 4$$

EXERCISE A

1. Given
$$\begin{array}{c|ccc}
a & 15 & 8 & 17 \\
b & 4 & 3 & 5 \\
c & 12 & 5 & 13
\end{array}$$

find A, B, C.

Solve the following triangles:

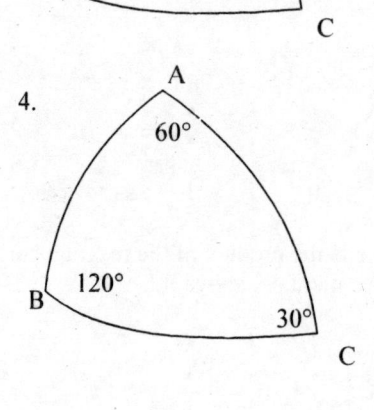

2.

5,12,13 A 5,12,13 4,3,5 B C

3.

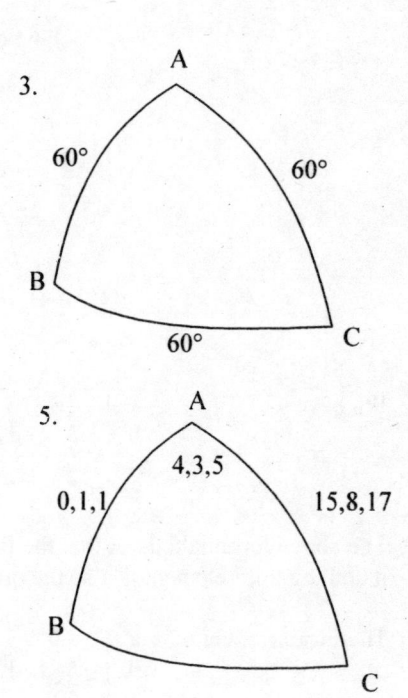

A 60° 60° B 60° C

4.

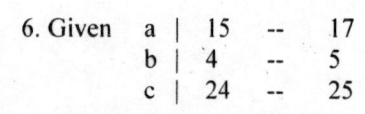

A 60° 120° B 30° C

5.

A 4,3,5 0,1,1 15,8,17 B C

6. Given
$$\begin{array}{c|ccc}
a & 15 & -- & 17 \\
b & 4 & -- & 5 \\
c & 24 & -- & 25
\end{array}$$

find A, B, C.

Further examples of spherical triangles which have to be solved by splitting them up into two right-angled triangles are given later.

6.2   RIGHT-ANGLED SPHERICAL TRIANGLES

There are ten formulae applicable to these triangles. We will give one of these, illustrate it, and then give the complete set. The right angle will be at C throughout.

Formula (1) is:

```
a   ●   ·   ✗        A   ·   ·   ·
b   ·   ·   |        B   ·   ·   ·
c   ●   ·   ✗        C   ·   ·   ·
```

EXAMPLE 3

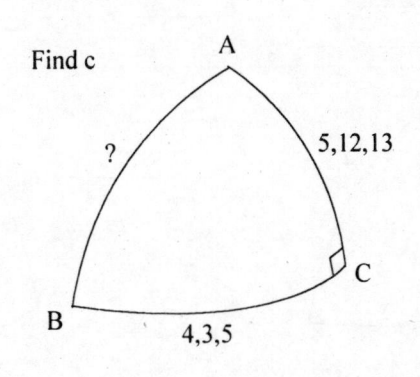

Find c

?

5,12,13

4,3,5

We have

```
a |  4   3   5        A |  ·   ·   ·
b |  5  12  13        B |  ·   ·   ·

c |  ·   ·   ·        C |  0   1   1    as C = 90°
```

The above formula tells us that the first element of c is the product of the two numbers above it and the third element is also the product of the two numbers above it.

Therefore, we get:

```
a |  4   3   5        A |  ·   ·   ·
b |  5  12  13        B |  ·   ·   ·
c | 20   -  65        C |  0   1   1
```

Or, dividing c by 5: c) 4, - , 13.

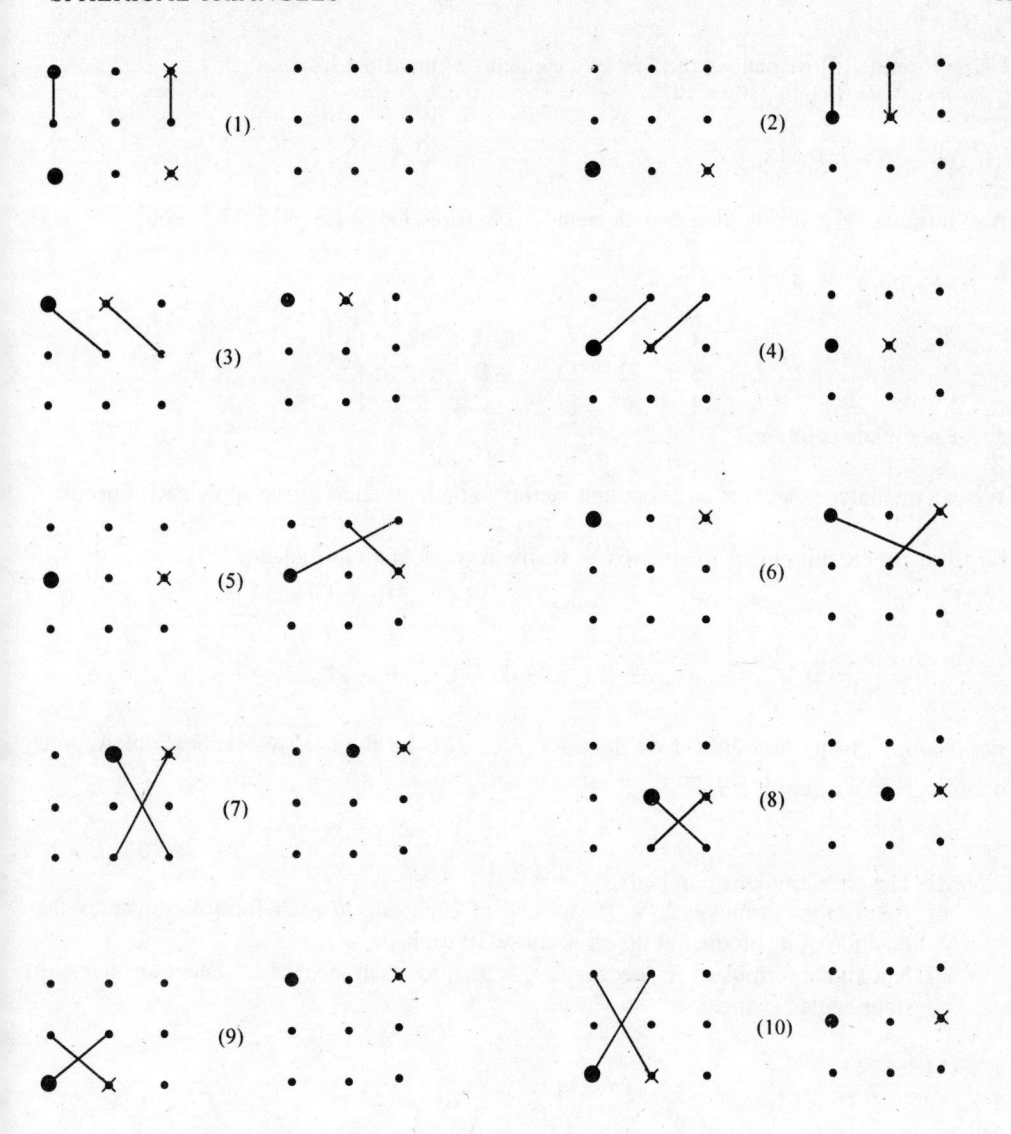

For proofs and the corresponding conventional formulae see Page 128.

We can now use some of these to complete the solution of the right-angled triangle on the previous page.

a	4	3	5	A	♦	♦	♦
b	5	12	13	B	♦	♦	♦
c	4	-	13	C	0	1	1

Using formula (3) we can get the first two elements of the triple for A.
That is: $4.12 = 48$ and $3.13 = 39$.

∴ A) 48, 39, - = 16, 13, -.

And formula (4) gives the first two elements of the triple for B: $5.3 = 15$, $12.5.= 60$.

∴ B) 15, 60, - = 1, 4, -.

So we have:

a	4	3	5	A	16	13	-
b	5	12	13	B	1	4	-
c	4	-	13	C	0	1	1

as the complete solution.

We can mentally cancel out any common factors to both products as we apply each formula.

By filling in the missing elements we can verify the remaining formulae:

a	4	3	5	A	16	13	$5\sqrt{17}$
b	5	12	13	B	1	4	$\sqrt{17}$
c	4	$3\sqrt{17}$	13	C	0	1	1

For example, using formula (7) we have 3.13, $5.3\sqrt{17}$ for the last two elements of A. And dividing by 3 we get 13, $5\sqrt{17}$.

Note: 1. The ten formulae go in pairs.
2. Apart from formulae 2, 9, 10, an end of each line in each formula indicates the position of its product in the adjacent set of triples.
3. Not all the formulae are needed: 1, 3, 4, 5, 6, for example, are sufficient to solve any right-angled triangle.

EXAMPLE 4

Solve:

a	12	5	13	A	8	15	17
b	•	•	•	B	•	•	•
c	•	•	•	C	0	1	1

Using (3) to get b we obtain b) - , $\frac{2}{3}$, 3 = - , 2, 9.

Using (7) to get c: c) - , $\frac{17}{13}$, 3 = - , 17, 39.

Using (6) to get B: B) - , $\frac{13}{17}$, $\frac{3}{2}$ = - , 26, 51.

There are, of course, other ways of arriving at the same results.

Therefore:

a	12	5	13	A	8	15	17	
b	-	2	9	B	-	26	51	
c	-	17	39	C	0	1	1	is the solution.

EXAMPLE 5

Solve:

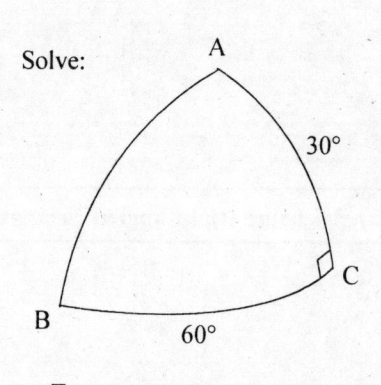

We have:

a | 1 √3 2   A | • • •
b | √3 1 2   B | • • •
c | • • •   C | 0 1 1

From (1) we get c) √3 , - , 4.
From (3) we get A) 1, 2√3 , -.
And from (4) we get B) 3, 2, -.

So

a | 1 √3 2   A | 1 2√3 -
b | √3 1 2   B | 3 2 -
c | √3 - 4   C | 0 1 1

EXERCISE B   Solve the following:

1. a | 4 3 5   A | • • •
    b | 12 5 13   B | • • •
    c | • • •   C | 0 1 1

2. a | 24 7 25   A | • • •
    b | • • •   B | • • •
    c | 3 4 5   C | 0 1 1

3. a | 4 3 5   A | • • •
    b | • • •   B | 12 5 13
    c | • • •   C | 0 1 1

4. a | • • •   A | 3 4 5
    b | • • •   B | 5 12 13
    c | • • •   C | 0 1 1

5. a | • • •   A | 12 5 13
    b | • • •   B | • • •
    c | 4 3 5   C | 0 1 1

6.
a	•	•	•		A	1	1	$\sqrt{2}$
b	•	•	•		B	0	1	1
c	•	•	•		C	0	1	1

Solution of Scalene Triangles using Right-angled Triangles

EXAMPLE 6

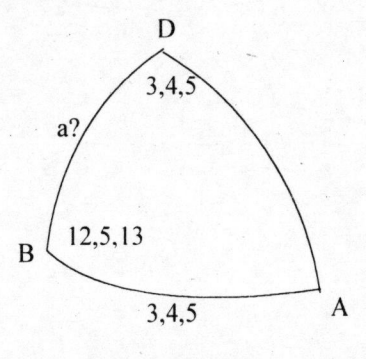

In this triangle we cannot find a using the Sine or the Cosine Rules or any of the other formulae. We could find AD using the Sine Rule but we could go no further.

However, if we split the triangle into two right-angled triangles as shown below, we can find a_1 and b from triangle ABC, then find a_2 from triangle ADC and add a_1 and a_2 to get a.

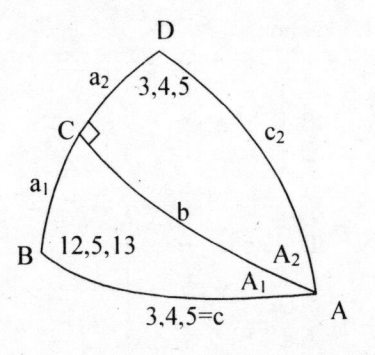

Triangle ABC:
a_1	•	•	•		A_2	•	•	•
b	•	•	•		B	12	5	13
c_1	3	4	5		C	0	1	1

From formula (10) we find a_1) 13, 16, -
then from (1) we get b) 12, 13, -.

Triangle ADC:
a_2	•	•	•		A_1	•	•	•
b	12	13	-		D	3	4	5
c_2	•	•	•		C	0	1	1

We find a_2 from formula (4): a_2) - , 13, 16.
Finally we use triple addition to add the angles:

a_1 |	13	16	$\sqrt{425}$	
a_2 |	$\sqrt{87}$	13	16	+

$$a = 13\sqrt{87} - 208, \quad 16\sqrt{87} + 169, \quad 16\sqrt{425}$$

EXAMPLE 7

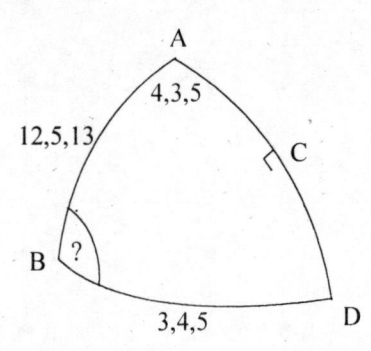

We can go through a similar procedure to find the angle B in this triangle. We find one part of B from triangle ABC, and also BC. Then we find the other part from triangle DBC and add the two parts.

6.3 SPHERICAL TRIANGLES USING CODE NUMBERS

In the solution of spherical triangles in the last chapter we saw how, given the triples for certain angles, we could find the triples for other angles. If the angles are given as triples the calculation is straightforward, but if they are given in radians or degrees we have to convert the angles into triples, find the triple of the answer and convert this back to an angle. And the code numbers are used to make these conversions. However, it could make the calculation easier, when the angles are in radians or degrees, to go straight from the code numbers of the given angles to the code numbers of the answer, thereby avoiding the triples themselves.

Let the code numbers of the sides a, b, c and the angles A, B, C of a spherical triangle be as follows:

a |	e	f		A |	E	F
b |	g	h		B |	G	H
c |	i	j		C |	I	J

This arrangement of angles and code numbers will be used throughout this chapter.

To find an Angle given Three Sides

If we have a, b, c and require A, the relation between the code numbers is:

$$E^2 = \{e(gj + hi)\}^2 - \{f(gi - hj)\}^2 \qquad F^2 = \{f(gi + hj)\}^2 - \{e(gj - hi)\}^2$$

Or diagrammatically:

$$E^2 =$$

$$F^2 =$$

The proof of this and the other results in this chapter are given on Pages 127-8.

EXAMPLE 8

Let us calculate again (see Page 70) the angle A in the triangle given by:

a	4	3	5
b	12	5	13
c	15	8	17

Converting to code numbers we have:

a	3	1
b	5	1
c	4	1

$$\therefore\ E^2 = \{3(5.1 + 1.4)\}^2 - \{1(5.4 - 1.1)\}^2 = 27^2 - 19^2 = 368.$$

$$F^2 = \{1(5.4 + 1.1)\}^2 - \{3(5.1 - 1.4)\}^2 = 21^2 - 3^2 = 432.$$

$$\therefore\ \frac{E}{F} = c(A) = \sqrt{\frac{368}{432}} = \sqrt{\frac{23}{27}}.$$

To confirm that this result is the same as that on Page 71, the triple for A is:

A) $E^2 - F^2$, 2EF, E^2+F^2

$\therefore$ A) 23–27, - , 23+27 = –2, - , 25 as before.

Given Two Sides and the Included Angle to find the Side Opposite

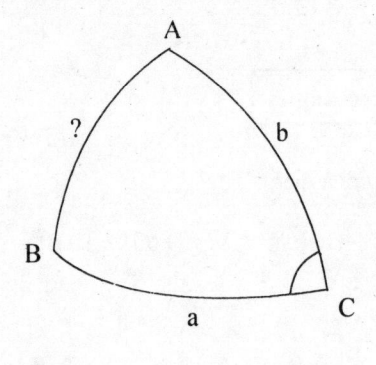

Suppose, we have a, b, C and require c.

Then: $i^2 = \{I(eg + fh)\}^2 + \{J(eg - fh)\}^2$ $j^2 = \{J(eh + fg)\}^2 + \{I(eh - fg)\}^2$

That is:

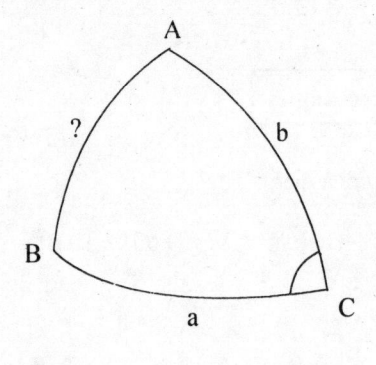

EXAMPLE 9

Given the code numbers: C | 4   3
                         a | 2   1
                         b | 3   2

find the code numbers i, j of c. (This is the same triangle as in the example on Page 71)

Using the patterns above: $i^2 = 32^2 + 12^2 = 1168$
$$j^2 = 21^2 + 4^2 = 457$$

So $\dfrac{i}{j} = c(c) = \sqrt{\dfrac{1168}{457}}$.

And the triple, *By Addition and By Subtraction*, is c) 711, - , 1625.

To find a Side given Three Angles

If we have A, B, C and require a then:

$$e^2 = \{E(GI + HJ)\}^2 - \{F(GJ - HI)\}^2 \qquad f^2 = \{F(GJ + HI)\}^2 - \{E(GI - HJ)\}^2$$

That is:

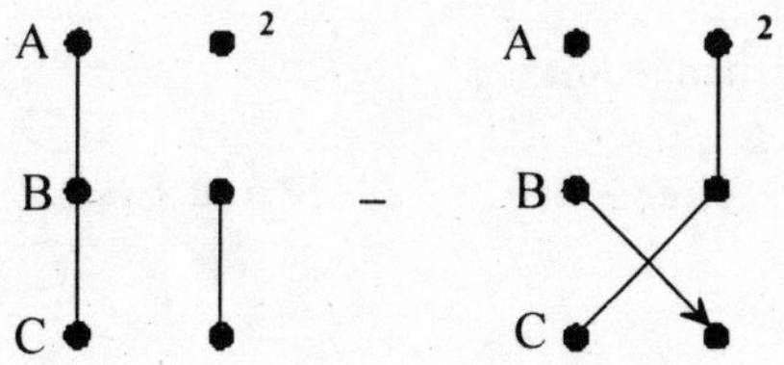

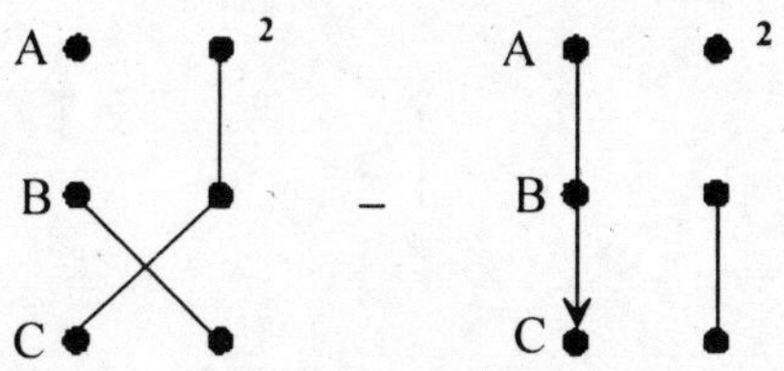

EXAMPLE 10

Given the code numbers
A	4	3
B	2	1
C	3	2

find the code numbers e, f of a.

From the patterns above: $e^2 = 32^2 - 3^2 = 1015$, $f^2 = 21^2 - 16^2 = 185$.

$\therefore \dfrac{e}{f} = c(a) = \sqrt{\dfrac{1015}{185}} = \sqrt{\dfrac{203}{37}}$.

Given Two Angles and the Side between them to find the Angle Opposite

---

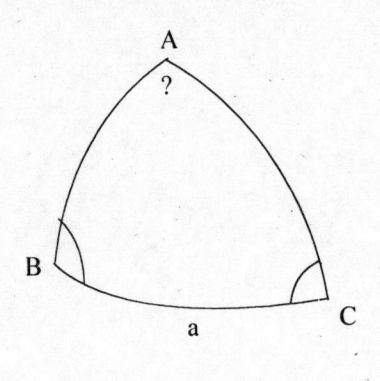

Here we have a, B, C and require A.

Then: $E^2 = \{e(GJ + HI)\}^2 + \{f(GJ - HI)\}^2$ and $F^2 = \{f(GI + HJ)\}^2 + \{e(GI - HJ)\}^2$

$$E^2 =$$

$$F^2 =$$

EXAMPLE 11

Given the code numbers: a | 5   1
                         B | 5   3
                         C | 2   1

find c(A).

We get: $E^2 = 55^2 + 1^2 = 3026$ and $F^2 = 13^2 + 35^2 = 1394$.

$\therefore \; \frac{E}{F} = c(A) = \sqrt{\frac{3026}{1394}} = \sqrt{\frac{1513}{697}}$.

Given Two Sides and an Angle Opposite to find the other Angle Opposite

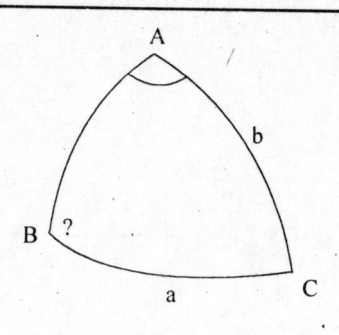

Given a, b, A we require B.

If the code numbers of a, b, A are:

$$\begin{array}{c|cc} a & e & f \\ b & g & h \\ A & E & F \end{array}$$

then G', H', the code numbers of the angle complementary to B, are given by:

$$G'^2 = (egE + fhF)(fgE + ehF) + (ehE + fgF)(fhE + egF) \text{ and}$$

$$H'^2 = (egE - fhF)(fgE - ehF) + (ehE - fgF)(fhE - egF).$$

We see here that the two expressions are the same except for the signs.

The pattern for this formula is shown below:

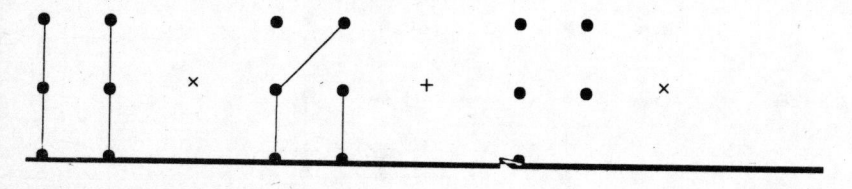

EXAMPLE 12

Given the code numbers:

$$\begin{array}{c|cc} a & 5 & 1 \\ b & 3 & 1 \\ A & 4 & 1 \end{array}$$

find B.

$$G'^2 = (60 + 1)(12 + 5) + (20 + 3)(4 + 15) = 1037 + 437 = 1474,$$
$$H'^2 = (60 - 1)(12 - 5) + (20 - 3)(4 - 15) = 413 - 187 = 226.$$

∴ the code number of the triple complementary to B is $\sqrt{\frac{737}{113}} = 2.5538$.

Then:

$$\begin{array}{cc} 2.554 & 1 \\ \underline{\quad 5 \quad} & \underline{\quad 2 \quad} \\ 14.77, & -0.108 \end{array} -$$

$\therefore$ B $= \frac{\pi}{2} - (0.7610 - \frac{0.216}{14.77}) = 0.8244$ radians.

Given Two Angles and a Side Opposite to find the other Side Opposite

---

EXAMPLE 13   Find a.

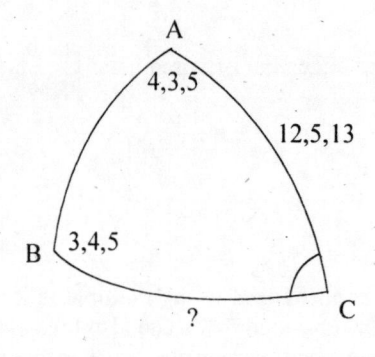

We have: B | 2   1
           A | 3   1
           b | 5   1

The pattern of the formula is the same as in the previous example.

$\therefore$ e$'^2$ = (30 + 1)(15 + 2) + (10 + 3)(5 + 6) = 670,

and f^2 = (30 − 1)(15 − 2) + (10 − 3)(5 − 6) = 370.

$\therefore$ c(a)(CT) $= \sqrt{\frac{67}{37}} = 1.3457$.

$$\begin{array}{cc} 1.3457 & 1 \\ \underline{4} & \underline{3} \quad - \\ 8.383, & -0.037 \end{array}$$

$\therefore$ $\frac{\pi}{2} - (\frac{\pi}{2} - 0.2838 - \frac{0.074}{8.383})$, as 4,3 = 1 − 7 in code number arithmetic.

$\therefore$ a = 0.2926 radians.

However, if in these last two examples the required angle is obtuse, the second quantity calculated, (egE − fhF)(fgE − ehF) + (ehE − fgF)(fhE − egF) in Example 12, will be negative so that this method cannot be used (since we cannot take the square root of a negative number). But this means that this quantity can be used to determine whether an angle is acute or obtuse.

6.4 DETERMINANTS

In the above examples the code numbers were conveniently small numbers and they are in fact always smaller than the elements of the triple that they represent. But if the code numbers are large, i.e. 2 or 3-figure numbers, the calculations can be lightened by the use of determinants.

A 2 by 2 determinant is an arrangement of 4 numbers: $\begin{vmatrix} a & b \\ c & d \end{vmatrix}$ whose value is found by cross-multiplying and subtracting.

Thus, $\begin{vmatrix} a & b \\ c & d \end{vmatrix} = ad - bc$.

The value of a determinant is unchanged if any multiple of a row or column is added to, or subtracted from, the other row or column. We can also take out any factor which is common to any row or column. Thus we can manipulate determinants to make some of the numbers smaller, to produce factors which can then be taken out or to obtain easy products.

EXAMPLE 14   Find the value of $\begin{vmatrix} 69 & -49 \\ 34 & 15 \end{vmatrix}$.

Adding the right-hand column to the left-hand column we get:

$$\begin{vmatrix} 69 & -49 \\ 34 & 15 \end{vmatrix} = \begin{vmatrix} 20 & -49 \\ 49 & 15 \end{vmatrix} = 300 + 2401 = 2701.$$

EXAMPLE 15   Find the value of $\begin{vmatrix} 69 & 49 \\ 15 & 34 \end{vmatrix}$.

We can take out a factor of 3 and then take twice the left-hand column from the right-hand column: $\begin{vmatrix} 69 & 49 \\ 15 & 34 \end{vmatrix} = 3\begin{vmatrix} 23 & 49 \\ 5 & 34 \end{vmatrix} = 3\begin{vmatrix} 23 & 3 \\ 5 & 24 \end{vmatrix} = 9\begin{vmatrix} 23 & 1 \\ 5 & 8 \end{vmatrix} = 9 \times 179 = 1611.$

EXAMPLE 16   Find the value of $\begin{vmatrix} 69 & -49 \\ 15 & 34 \end{vmatrix}$.

$$\begin{vmatrix} 69 & -49 \\ 15 & 34 \end{vmatrix} = 3\begin{vmatrix} 23 & -49 \\ 5 & 34 \end{vmatrix} = \begin{vmatrix} 23 & -26 \\ 5 & 39 \end{vmatrix} = 39\begin{vmatrix} 23 & -2 \\ 5 & 3 \end{vmatrix} = 39 \times 79 = 3081.$$

EXAMPLE 17   Find the value of $\begin{vmatrix} 69 & 49 \\ 34 & 15 \end{vmatrix}$.

$$\begin{vmatrix} 69 & 49 \\ 34 & 15 \end{vmatrix} = \begin{vmatrix} 1 & 19 \\ 34 & 15 \end{vmatrix} = \begin{vmatrix} 1 & 20 \\ 34 & 49 \end{vmatrix} = 49 - 680 = -631.$$

Application of Determinants

EXAMPLE 18   Given a spherical triangle with a = 0.6435, b = 1.235, c = 0.831 radians, find A.

Code numbers for a are 3,1,
Code numbers for b are 69, 49 (see "Triples" Page 98),
Code numbers for c are 34,15.

This is similar to Example 1 in this chapter, so we have:

$$c^2(A) = \frac{\left\{3(69 \times 15 + 49 \times 34)\right\}^2 - \left\{1(69 \times 34 - 49 \times 15)\right\}^2}{\left\{1(69 \times 34 + 49 \times 15)\right\}^2 - \left\{3(69 \times 15 - 49 \times 34)\right\}^2}.$$

Or, $c^2(A) = \dfrac{\left\{3\begin{vmatrix} 69 & -49 \\ 34 & 15 \end{vmatrix}\right\}^2 - \left\{1\begin{vmatrix} 69 & 49 \\ 15 & 34 \end{vmatrix}\right\}^2}{\left\{1\begin{vmatrix} 69 & -49 \\ 15 & 34 \end{vmatrix}\right\}^2 - \left\{3\begin{vmatrix} 69 & 49 \\ 34 & 15 \end{vmatrix}\right\}^2}.$

But these are the determinants evaluated in Examples 14-17.

So $c^2(A) = \frac{8103^2 - 1611^2}{3081^2 - 1893^2} = \frac{65660 - 2595}{9493 - 3583} = \frac{63065}{5910} = 10.6709.$

In second step above we have made use of the simple Vedic one-line method of squaring numbers from the left and in the last step we have used the Vedic one-line method of division shown in Chapter 1. Next we take the square root, again by the Vedic one-line method.

$\therefore\ c(A) = 3.2666.$

Finally we can decode to get A. The calculation is similar to that in Example 10 Page 52.

$$A = 0.6435 + \frac{2(3 - 3.2666)}{3 \times 3.2666 + 1} = 0.5941 \text{ radians.}$$

Determinants are also useful in the addition and subtraction of triples.

EXAMPLE 19 Find 36   77   85
                   56   33   65 +

The answer is $\begin{vmatrix} 36 & 77 \\ 33 & 56 \end{vmatrix}, \begin{vmatrix} 36 & -77 \\ 56 & 33 \end{vmatrix}$, 85×65.

$$\begin{vmatrix} 36 & 77 \\ 33 & 56 \end{vmatrix} = 3 \times 7 \begin{vmatrix} 12 & 11 \\ 11 & 8 \end{vmatrix} = 3 \times 7 \begin{vmatrix} 1 & 3 \\ 11 & 8 \end{vmatrix} = 3 \times 7(-25) = -525.$$

$$\begin{vmatrix} 36 & -77 \\ 56 & 33 \end{vmatrix} = 4 \times 11 \begin{vmatrix} 9 & -7 \\ 14 & 3 \end{vmatrix} = 44 \times 125 = 5500.$$

So the total of the two triples is $-525, 5500, 85 \times 65 = -21, 220, 221$.

The Cotangent Rule

If we have two sides and the angle between them in a spherical triangle, and we want another angle, we can use the cotangent rule, as described on Page 75 where we found a) 956, 255, - , and since 956, 255 are the code numbers of the angle 2a we can find 2a and then a:

  956 255
     4   1 –
 4079 64 $\therefore 2a = 0.4900 + \frac{128}{4079} = 0.52134.$

$\therefore a = 0.2607.$

On Pages 75-6 we found A) 191, 612, -. If we take 191, 612 as code numbers we see that they represent an obtuse angle.

We have a choice:

(1)   612, 191 are the code numbers of the supplementary triple:

  612 191
     3   1 –
 2027 –39 $\therefore 2A = \pi - (0.6435 - \frac{78}{2079})$

$\therefore A = 1.2683$ radians.

(2) 803, 421 (by addition and by subtraction) are the code numbers of the complementary triple:

803   421
  2   1 −
2027 39 $\therefore$ $2A = \frac{\pi}{2} + (0.9273 + \frac{78}{2027})$

$\therefore$ $A = 1.2683$ radians.

6.5 SUMMARY

The diagrams used in this chapter show that there are simple *Vertical and Crosswise* patterns behind the otherwise complex-looking formulae. These patterns could be used to make computer programmes run more efficiently.

Since also any angle can be represented by a perfect triple and the angle in a perfect triple can be found to any desired degree of accuracy, the methods shown here have a general application to the solution of spherical triangles. But with the widespread us of calculators and computers nowadays these would not usually be appropriate techniques. However, they can be used to easily give approximate answers and provide checks. The methods being simple they also provide an elementary introduction to the subject of spherical triangles and their solution. And there is something more satisfying about obtaining an exact solution using a simple method.

Chapter 7

QUADRUPLES

7.1 INTRODUCTION

A 3-dimensional equivalent to triples can be used to define a direction in 3-dimensional space, just as ordinary triples can define any direction in 2-dimensional space. This leads to the notion of "quadruples".

These can be defined and developed along similar lines to the triples and we will see that they have useful applications in astronomy.

6, 3, 2, 7 is a **perfect quadruple** as $6^2 + 3^2 + 2^2 = 7^2$ $(36 + 9 + 4 = 49)$.

Any four numbers that have this property is a quadruple, and if the numbers are also rational as above it is a perfect quadruple.

So if x, y, z, r is a quadruple then $x^2 + y^2 + z^2 = r^2$, and conversely.

The quadruple 6, 3, 2, 7 describes the point P(6, 3, 2) in 3-dimensional space which is exactly 7 units from the origin O:

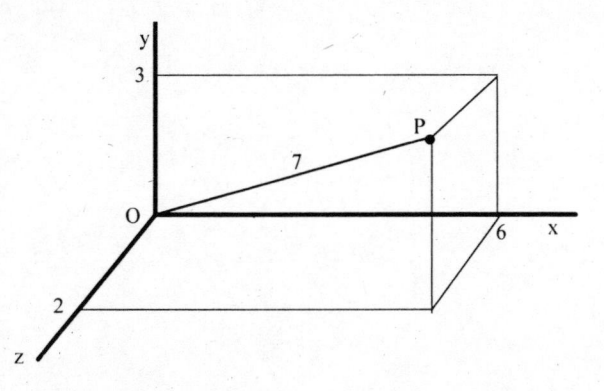

Other examples of perfect quadruples are:

1	70	70	99
2	−1	2	3
0	3	4	5
49	64	8	81

Note that the third example here: 0,3,4,5 is just a triple in the y-z plane.

If any element of a quadruple is zero the quadruple is also a triple, so all triples are contained within the 3-dimensional quadruples.

Quadruples can be represented usefully by a pyramid composed of 4 right-angled triangles:

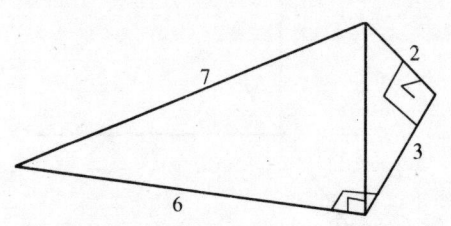

We will discuss later what we mean by the angle in a quadruple.

The formula $c^2 - d^2 + e^2$, $2cd$, $2de$, $c^2 + d^2 + e^2$ generates perfect quadruples for all rational values of c,d,e.

c, d, e are the **code numbers** of the quadruple.

For example, for code numbers **3, 1, 2** the above formula gives **12, 6, 4, 14 = 6, 3, 2, 7**.

There are a number of other formulae that also generate perfect quadruples. And in addition perfect quadruples can be produced:

    (a) from triples (compare the generating formulae),
    (b) by the addition of two triples (see Pages 101 and 106),
    (c) by the addition of a triple and a quadruple (see Page 103),
    (d) by the addition or subtraction of two quadruples (see Pages 106-7).

The code numbers of the quadruple **6, 3, 2, 7** are **3, 7–6, 2 = 3, 1, 2**.
That is, in general, the code numbers of **x, y, z, r** are **y, r–x, z**.

You may like to check that the code numbers of the four quadruples given on Page 130 are:
                5, 7, 5, (any common factor can be divided out)
          –1, 1, 2,
          3, 5, 4
   and   **8, 4, 1.**

So the code numbers **1, 1, 3** generate the quadruple **9, 2, 6, 11**,
and the code numbers of **9, 2, 6, 11** are **2, 2, 6 = 1, 1, 3**.

Similarly code numbers **c,d,e** generate $c^2 - d^2 + e^2$, $2cd$, $2de$, $c^2 + d^2 + e^2$.
And the code numbers of $c^2 - d^2 + e^2$, $2cd$, $2de$, $c^2 + d^2 + e^2$ are $2cd$, $2d^2$, $2de$ = **c, d, e**.

The quadruple **1, 0, 0, 1** describes the positive x-axis.
The six fundamental directions are given below with their code numbers:

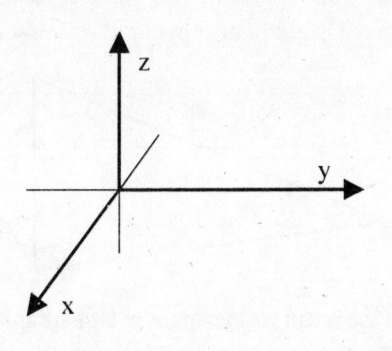

	quadruple	code numbers
+ve x-axis	1 0 0 1	1 0 1*
−ve x-axis	−1 0 0 1	0 1 0
+ve y-axis	0 1 0 1	1 1 0
−ve y-axis	0 −1 0 1	1 −1 0
+ve z-axis	0 0 1 1	0 1 1
−ve z-axis	0 0 −1 1	0 −1 1

* For this quadruple the code numbers cannot be found by the method described in the last section. It is the only exception however.

We define quadruple subtraction as follows:

$$\frac{\begin{array}{cccc} x & y & z & r \\ X & Y & Z & R \end{array}}{xX + yY + zZ, \quad yX - xY, \quad zX - xZ, \quad zY - yZ, \quad rR} \quad - \qquad \dots (1)$$

For example:

$$\frac{\begin{array}{cccccc} 2 & 3 & 6 & & 7 \\ 2 & 2 & 1 & & 3 \end{array}}{16 \quad 2 \quad 10 \quad 9 \quad 21} \quad -$$

The result is a "quintuple": $16^2 + 2^2 + 10^2 + 9^2 = 21^2$.
That the difference of any two quadruples is a quintuple (that is, that the sum of the squares of the first four elements is equal to the square of the 5th) is easily proved from (1) above.

We define quadruple subtraction in this way because:

(a) if one element, say the 4th, in (1) is zero (i.e. if zY = yZ) we obtain a perfect quadruple;
as will be seen later this leads to very interesting and useful applications,

(b) if the middle three elements in (1) are combined into one element the result is a triple with useful properties (see Section 7.6).

7.2   ADDITION OF PERPENDICULAR TRIPLES

In astronomy positions in space are often described by an angle measured from some reference direction in a reference plane, and another angle perpendicular to this plane:

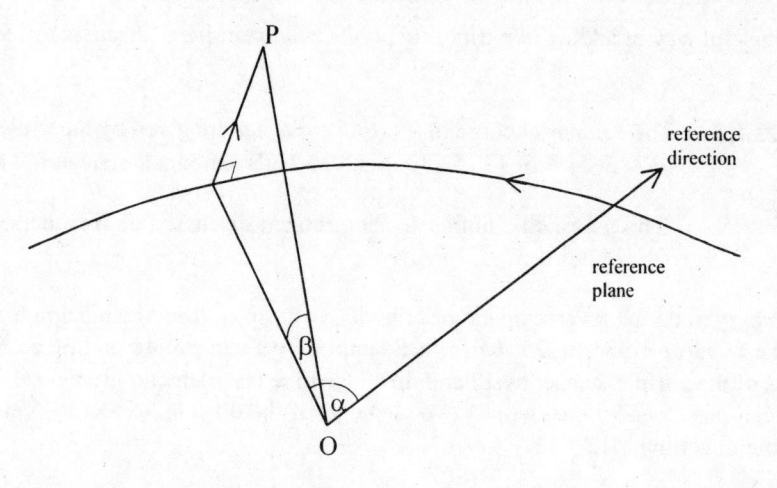

So the position of P is described by two perpendicular angles α, β.

But since a position in space can also be described by a quadruple, there must be a relationship between the elements of the triples for α, β and the corresponding quadruple for the same position.

Suppose, that the two triples are α) **3 4 5** β) **8 15 17** then the equivalent quadruple is **24 32 75 85**.

For 24 and 32 we multiply the first two elements of the first triple by the first element of the second triple, and for 75 and 85 we multiply the last two elements of the last triple by the last element of the first triple.

This is an application of the Vedic formula *The First by the First and the Last by the Last*. We can see why this procedure works if we place the 8 side of the β triple along OQ below, which is the hypotenuse of the α triple:

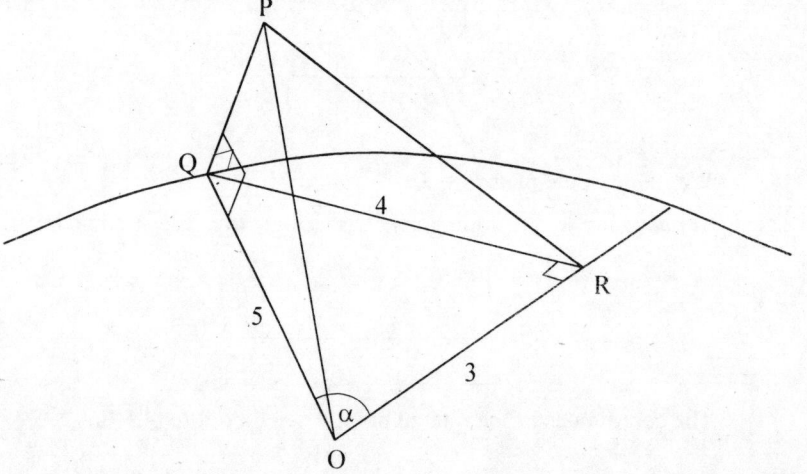

Using *Proportion* we take the β triple as $5, \frac{5 \times 15}{8}, \frac{5 \times 17}{8}$ to make the two triples "fit".

Then we can see that the quadruple is $3, 4, \frac{5 \times 15}{8}, \frac{5 \times 17}{8} = 24, 32, 75, 85$.

Thus two perfect triples produce a perfect quadruple.

Another useful way of adding two triples to produce a quadruple is discussed in Section 5.

EXAMPLE 1   The equatorial coordinates (α, δ) of a star are given by the triples
                α) 4 3 5,   δ) 12 5 13. Find the star's rectangular equatorial coordinates.

                This is exactly similar to the problem discussed above: the answer is 48,
                36, 25, 65.

It is also easy to do the reverse operation: find the two triples from the quadruple.
Given the quadruple 48, 36, 25, 65 from Example 1 we can put 48 and 36 as the first two
elements of the a triple, cancel by 12 and then insert the last element, giving α)4 3 5;
and we can put 25 and 65 as the last two elements of the d triple, cancel by 5 and insert the
first element, getting δ)12 5 13.

7.3   ROTATION ABOUT A COORDINATE AXIS

See page 41 for triple rotations. When rotating a point about, say, the x-axis, the x-coordinate
of the point will not change and the rotation is in a plane parallel to the y, z plane.

EXAMPLE 2   Rotate the point P(4, 2, 3) about the x-axis by the angle 4, 3, 5.

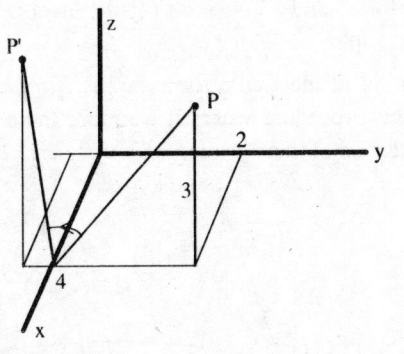

We rotate in the plane x = 4.

In this plane we are adding a 4, 3, 5 triangle to a 2, 3, - triangle:

$$
\begin{array}{cccc}
4 & 2 & 3 & \sqrt{13} \\
  & 4 & 3 & \,\underline{5} \quad + \\
\hline
  & -\frac{1}{5} & \frac{18}{5} & \sqrt{13}
\end{array}
$$

The coordinates of the rotated point P' are therefore (4, –0.2, 3.6).

EXAMPLE 3 Rotate P(4, 2, 3) by angle 4, 3, 5 about a) the y-axis   b) the z-axis.

(a) The y coordinate of P is unchanged.   (b) The z-coordinate is unchanged.

4	2	3	-		4	2	3	-
4		3	5 +		4	3		5 +
$\frac{7}{5}$	2	$\frac{24}{5}$			$\frac{10}{5}$	$\frac{20}{5}$	3	

∴ P'(1.4, 2, 4.8).   ∴ P'(2, 4, 3).

We do not need to calculate the hypotenuse for P each time and since we invariably divide through by the third element of the rotation triple we can disregard the 4th column and write the answer straight down.

Note that for rotations about the y-axis, adding the triples rotates clockwise from the positive y-direction (but anticlockwise for the other two axes).

We can also use this method to rotate shapes in 3-dimensional space (about an inclined axis) by making small rotations about two or three of the coordinate axes in turn.

Change of Coordinate System

Here we use the method from the last chapter for rotating about a coordinate axis.

If we have the coordinates of a point P(x', y', z') in a rectangular coordinate system x', y', z' and we want its coordinates in another rectangular coordinate system x, y, z with the same origin, we can get them by using the triple rotation method described in the last section.

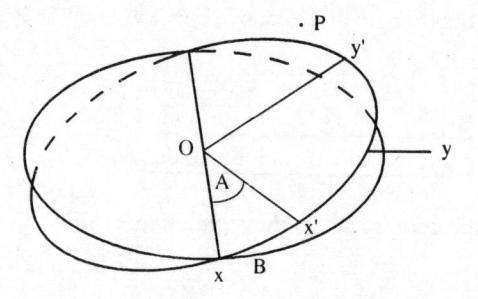

Suppose that the x-y plane of these two systems are inclined at B)l, m, n, and let us suppose for the moment that Ox is along the line of intersection of these planes, as shown above.

Let the angle between Ox and Ox' be given by A)i, j, k.

So A and B describe the orientation of the two frames of reference and we suppose that these, and P(x', y', z') are known. We want P(x, y, z).

Consider the xx'y' plane viewed from the z'O direction:

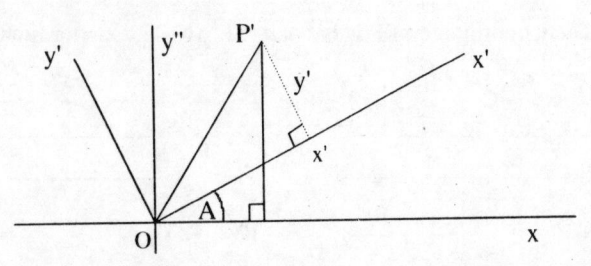

P' is the projection of P on this plane.

We are effectively rotating x'Oy' about the Oz' axis by the angle A so that Ox' coincides with Ox and Oy' falls on Oy".

$$
\begin{array}{c|cccc}
 & x' & y' & z' & r \\
A & i & j & k & \quad + \\
\hline
 & x'i-y'j & y'i+x'j & z'k & rk \quad = a, b, c, d \text{ say.}
\end{array}
$$

Next we rotate this axis system xy"z' about the x-axis by the angle B:

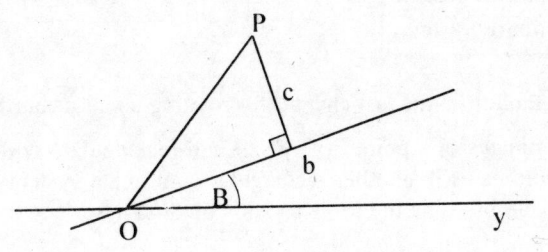

Note that a corner of the triangle drawn above is not at O but at a point a on the x-axis. Adding we get:

$$
\begin{array}{c|cccc}
 & a & b & c & d \\
B & & l & m & n \quad + \\
\hline
 & an & nl-cm & cl+bm & dn \quad = A, B, C, D \text{ say.}
\end{array}
$$

EXAMPLE 4 Find the coordinates in the x, y, z frame of P(2,2,1) given A)4 3 5 and
B)12 5 13.

$$
\begin{array}{c|cccc}
 & 2 & 2 & 1 & 3 \\
A & 4 & 3 & 5 & \quad + \\
\hline
 & 2 & 14 & 5 & 15 \\
B & & 12 & 5 & 13 \quad + \\
\hline
 & 26 & 143 & 130 & 195 \\
= & 2 & 11 & 10 & 15 \\
= & 0.4 & 2.2 & 2 & 3
\end{array}
$$

So P has coordinates (0.4, 2.2, 2) in the other frame.

If, however the Ox axis is not along the line of intersection of the xy and x'y' planes a further rotation is needed: about Oz.

Suppose C)p,q,s is the angle between Ox and the line of intersection of the planes:

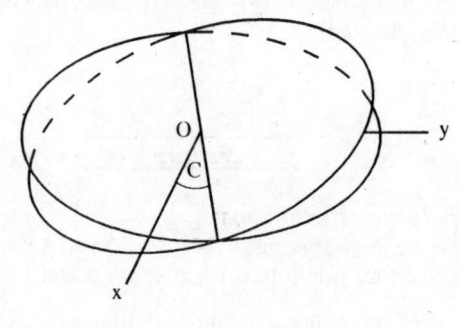

P will be given by:

A	B	C	D
p	q	s	+
Ap-Bq	Bp+Aq	cs	Ds

Clearly this method could be extended to 3 or more inclined planes.
Note also that the quadruple for P could represent a third orbit plane rather than the position of a point.

7.4   QUADRUPLES AND ORBITS

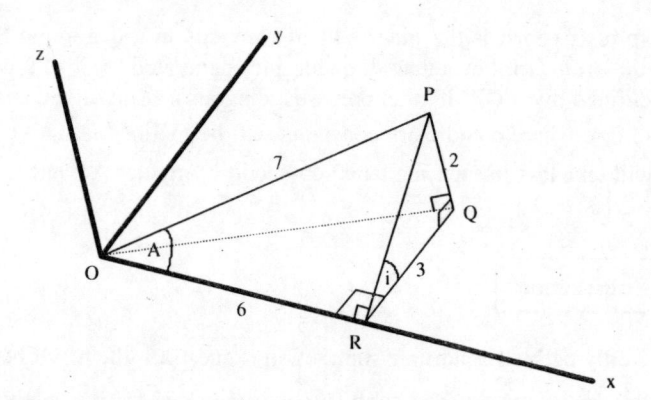

The diagram shows the quadruple 6, 3, 2, 7.
Angle POR = A, i.e. $A = \cos^{-1}\left(\frac{6}{7}\right)$.

Angle PRQ = i, i.e. $i = \tan^{-1}\left(\frac{2}{3}\right)$.

A quadruple can be used to represent an orbit crossing, in this case, the x-axis.

i is the inclination of the orbit to the reference frame and the angle A defines a particular position in the orbit.

Quadruple for a given i and A

It is sometimes necessary to know the quadruple which has a given inclination, i, and argument, A. Suppose, for example, that we have i)12 5 13 and A) 4 3 5 and that we want to know the quadruple that they define:

$$
\begin{array}{ccc}
12 & 5 & 13 \\
4 & 3 & 5 \\
\hline
52 \quad 36 & 15 & 65
\end{array}
$$

We use the *Vertically and Crosswise* formula:
> for 52 and 36 we multiply cross-wise, 4×13 and 12×3,
> and for 15 and 65 we multiply vertically, 5×3 and 13×5.

This result is easily proved by using a method similar to that used in Section 1 for the addition of two perpendicular triples.

Inclination of Orbit

In the quadruple 6, 3, 2, 7 the inclination, i, is determined by the two inner elements 3, 2 and is independent of the values of the outer elements- which determine A (see Section 1).
Consider the general quadruple $c^2-d^2+e^2$, 2cd, 2de, $c^2+d^2+e^2$ generated by the code numbers c, d, e. The ratio of the inner elements of the quadruple is the same as the ratio of the outer elements of the code numbers.
So all quadruples generated by, say, 3, d, 2 will be of the form x, 3, 2, r for all values of d.

But as we have just seen it is the inner pair of elements in a quadruple that determine the inclination of the orbit. It follows that all quadruples generated by 3, d, 2 will be in the same orbit plane (defined by POR in the previous diagram). And in general all quadruples generated by c, d, e where c and e are constants will be inclined at $\tan^{-1}\left(\frac{e}{c}\right)$ to the xy plane for all d. We will take this further in a later sub-section: Angular Advance.

Quadruple Subtraction

So far we have only defined quadruple subtraction (Page 100), the result being a quintuple.

We would like to be able to add and subtract quadruples and for the result to be a quadruple. And, of course, we would like the process to serve some useful purpose.

Let us impose the condition (to hold in the following sections) that when adding or subtracting two quadruples the ratio of their inner elements is the same. The geometrical consequence of this, as discussed in the last section, is that the two quadruples will be in the same orbit. Furthermore this will result in one of the elements of the quintuple being zero: thereby giving a quadruple rather than a quintuple.

For example, suppose we want to find $\dot{2},6,3,7 - 2,2,1,3$ in which the ratio of the inner elements is the same, i.e. $6:3 = 2:1$. The subtraction gives (see Page 100):

$$
\begin{array}{ccccc}
2 & 6 & 3 & 7 & \\
\underline{2} & \underline{2} & \underline{1} & \underline{3} & - \\
19 & 8 & 4 & 0 \quad 21 &
\end{array}
\qquad \ldots (2)
$$

which has a zero element and so can be compressed into the quadruple 19, 8, 4, 21.

Note that the inner elements of this quadruple are in the same ratio as the two subtracted quadruples: $8:4 = 2:1$, so it is in the same plane also. And note the operation of the Vedic formula *One in Ratio, the Other One Zero*.

Quadruple Addition

For subtraction we have:

$$
\begin{array}{cccc}
x & y & z & r \\
\underline{X} & \underline{Y} & \underline{Z} & \underline{R} \quad - \\
xX + yY + zZ, & yX - xY, & zX - xZ, & rR
\end{array}
\qquad \ldots (3)
$$

For addition we will have:

$$
\begin{array}{cccc}
x & y & z & r \\
\underline{X} & \underline{Y} & \underline{Z} & \underline{R} \quad + \\
xX - yY - zZ, & yX + xY, & zX + xZ, & rR
\end{array}
\qquad \ldots (4)
$$

in which the patterns are all the same and the signs are all reversed.

That this has the desired effect of reversing the subtraction can be verified by taking the subtraction sum:

$$
\begin{array}{cccc}
x & y & z & r \\
\underline{X} & \underline{ny} & \underline{nz} & \underline{R} \quad - \\
xX + ny^2 + nz^2, & yX - nxy, & zX - nxz, & rR
\end{array}
$$

and adding the last two lines as indicated in (4).

On cancelling and dividing through by $X^2 + n^2y^2 + n^2z^2$ (=R^2) the result should be x,y,z,r thus verifying that (4) is a legitimate addition process.

So let us now add 19, 8, 4, 21 and 2, 2, 1, 3 using this addition algorithm:

$$
\begin{array}{cccc}
19 & 8 & 4 & 21 \\
\underline{2} & \underline{2} & \underline{1} & \underline{3} \quad + \\
18 & 54 & 27 & 63 \quad = \quad 2, 6, 3, 7
\end{array}
$$

EXAMPLE 5

$$
\begin{array}{cccc}
1 & 2 & 2 & 3 \\
1 & 2 & 2 & 3 \ + \\
\hline
-7 & 4 & 4 & 9
\end{array}
\qquad \text{and} \qquad
\begin{array}{cccc}
-7 & 4 & 4 & 9 \\
1 & 2 & 2 & 3 \ - \\
\hline
9 & 18 & 18 & 27
\end{array}
= 1, 2, 2, 3.
$$

Doubling and Halving a Quadruple

Doubling a quadruple follows directly from the last section. For equal quadruples the addition process given in (4) yields:

$$2(x,y,z,r) = x^2 - y^2 - z^2, \ 2xy, \ 2xz, \ r^2.$$

So $2(7,4,4,9) = 17, 56, 56, 81$.

The formula for halving a quadruple is:

$$\tfrac{1}{2}(x,y,z,r) = x + r, \ y, \ z, \ (\sqrt{r}) \quad \text{or} \quad -(x+r), \ -y, \ -z, \ (\sqrt{r}).$$

The first three elements define the result. $\sqrt{r}$ should only be inserted for the fourth element when x,y,z,r is a perfect quadruple and the first three elements have been cancelled down.

So $\tfrac{1}{2}(17,56,56,81) = 98, 56, 56, (\sqrt{81}) = 7, 4, 4, 9$ or $-7, -4, -4, 9$.

The geometrical proof of this halving process follows very similar lines to that for halving the angle in a triple. (see Chapter 4).

Code Number Addition and Subtraction

For triples it was found to be a great advantage to be able to add and subtract the triple code numbers rather than the triples themselves, because we often work with the code numbers or require the code number of a final result rather than the triple itself. It is the same with the quadruples, we would like to be able to add and subtract the code numbers as well as the quadruples.

Let c,d,e and c,D,e be the code numbers of two quadruples which we want to add (we can use c,D,e rather than nc,D,ne because we can divide code numbers through by any constant without changing the quadruple it generates).

If we obtain the quadruples that these code numbers generate (as shown in Section 7.1), add the quadruples (as shown by (4) on Page 107), obtain the code numbers of the result (as shown on Page 97), simplify and divide through by $2(d + D)$ we obtain:

$$
\begin{array}{ccc}
 & d & \\
c & & e \quad + \\
 & D & \\
\hline
c(c^2+e^2-dD), & (c^2+e^2)(d+D), & e(c^2+e^2-dD)
\end{array}
$$

for addition and:

$$\begin{array}{c} \mathbf{d} \\ \mathbf{c} \qquad\qquad\qquad\qquad \mathbf{e} \qquad\qquad - \\ \mathbf{D} \\ \hline c(c^2+e^2+dD), \quad (c^2+e^2)(d-D), \quad e(c^2+e^2+dD) \end{array}$$

for subtraction. Note that because c and e are the same in both code numbers we need write them only once, so that the numbers form a diamond pattern.

EXAMPLE 6   Find the code numbers of the quadruple which is the sum of the quadruples whose code numbers are 3, 1, 2 and 3, 3, 2.

Using the above formula:

$$\begin{array}{c} 1 \\ 3 \qquad\qquad\qquad 2 \;+ \\ 3 \\ \hline 30 \qquad 52 \qquad 20 \quad = \quad 15, 26, 10 \end{array}$$

We can of course also do this the long way: find the quadruples from the code numbers, add them and find the code numbers of the result. The quadruples generated by 3,1,2 and 3,3,2 are 6, 3, 2, 7 and 2, 9, 6, 11 and their sum is –27, 60, 40, 77. The code numbers of this quadruple are 15, 26, 10, as before.

Angle in a Quadruple

The angle in a quadruple is the angle A in the diagram at the beginning of Section 7.4.

We now ask, what is the relationship between the angles in two quadruples and the angle in their sum or difference?

In fact the addition and subtraction processes defined here have the effect of adding and subtracting the angles respectively.

For example, let the angles in the quadruples in (2) be A, B, C.-

A	2	6	3	7
B	2	2	1	3	–
C	19	8	4	21

Then A – B = C, or $\cos^{-1}(\frac{2}{7}) - \cos^{-1}(\frac{2}{3}) = \cos^{-1}(\frac{19}{21})$.

We can easily verify this by triple subtraction:

$$\begin{array}{c} \cos^{-1}(\frac{2}{7})| \qquad 2 \qquad \sqrt{45} \qquad 7 \\ \cos^{-1}(\frac{2}{3})\,| \qquad 2 \qquad \sqrt{5} \qquad 3 \qquad - \\ \hline \qquad\quad | \; 4+\sqrt{45\times5}, \quad -, \quad 21 \end{array}$$

and the angle in 19, - , 21 is $\cos^{-1}(\frac{19}{21})$

Let us now add the quadruples 1, 2, 2, 3 and 7, 4, 4, 9:

$$
\begin{array}{c|cccc}
D & 1 & 2 & 2 & 3 \\
E & 7 & 4 & 4 & 9 \quad + \\
\hline
F & -9 & 18 & 18 & 27 = -1, 2, 2, 3
\end{array}
$$

And adding the angles in D and E:

The general result is easily proved algebraically.

$$
\begin{array}{c|ccc}
\cos^{-1}\left(\tfrac{1}{3}\right) & 1 & \sqrt{8} & 3 \\
\cos^{-1}\left(\tfrac{7}{9}\right) & 7 & \sqrt{32} & 9 \quad + \\
\hline
 & 7 - \sqrt{8 \times 32}, & -, & 27
\end{array}
$$

and the angle in $-1, -, 3$ is $\cos^{-1}\left(-\tfrac{1}{3}\right)$.

So D + E = F, and adding the quadruples adds the angles in the quadruples.

It appears then that this addition and subtraction method for quadruples would have useful astronomical applications: when, for example, a body in an inclined orbit (inclined to some reference plane) advances in its orbit by a certain amount, and we want to know its new position relative to the same reference plane.

Angular Advance

Now we know (see "Inclination of Orbit" earlier) that quadruples generated by code numbers c,d,e, where c and e are given, are in the same plane for all values of d. For example, consider the code numbers 2,d,1 for various values of d:

code numbers			generated quadruple			
2	0	1	1	0	0	1
2	1	1	2	2	1	3
2	2	1	1	8	4	9
2	3	1	-2	6	3	7
2	4	1	-11	16	8	21
etc.			etc.			

Note: 1. whole number values of d have been selected here, but any values may be chosen,
  2. as d increases, the angle in the generated quadruple increases,
  3. the first element of the 4th and 5th quadruples is negative, indicating that the angles are greater than 90°.

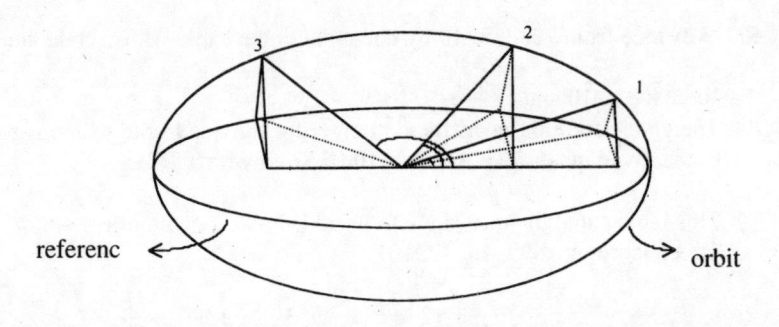

The diagram above shows three quadruples with increasing angles and their relationship to the reference plane.

In fact as d increases from 0 to ∞ (infinity), A increases from 0° to 180°.
Quadruples with arguments greater than 180° are produced by code numbers with negative d values: as d increases from −∞ to 0, A increases from 180° to 360°.

The angular advance as d increases from 0 to 1, from 1 to 2 etc. can be found by subtracting the quadruples:

$$
\begin{array}{cccc}
2 & 2 & 1 & 3 \\
1 & 0 & 0 & 1 \\
\hline
2 & 2 & 1 & 3
\end{array}
\quad -
\qquad\qquad
\begin{array}{cccc}
1 & 8 & 4 & 9 \\
2 & 2 & 1 & 3 \\
\hline
22 & 14 & 7 & 27
\end{array}
\quad -
$$

The angular advance in these two cases is $\cos^{-1}(\frac{2}{3})$ and $\cos^{-1}(\frac{22}{27})$.
It is worth noting that the code numbers of 22,14,7,27 are 14,5,7, so that the d number of this quadruple is $\frac{5}{7}$.

Now we may want to add a quadruple to another quadruple so that the added quadruple increases the angle by a certain given amount.

Relationship between d and A

This relationship is expressed by $c^2 + e^2 = (kd)^2$, where **k** is the triple code number of A.
So c, e, kd is a triple.

This is easily proved as follows. Code numbers c, d, e generate the quadruple
$c^2-d^2+e^2$, 2cd, 2de, $c^2+d^2+e^2$. This contains the angle given by the triple $c^2-d^2+e^2$, - , $c^2+d^2+e^2$.
The code number of this triple (see "Triples" Page 57, Example 8) is given by

$$
k = \sqrt{\frac{(c^2+e^2)}{d^2}} = \frac{\sqrt{c^2+e^2}}{d}.
$$

So $c^2 + e^2 = (kd)^2$.

EXAMPLE 7 Advance from 12, 4, 3, 13 by the angle in the triple whose code number is 10.

We have k=10 and c=4, e=3.

The above formula gives the d number for the quadruple which is to be added to the given quadruple: $4^2+3^2 = (10d)^2$, from which $d=\frac{1}{2}$.

This means that the quadruple to be added has code numbers 4, ½, 3 = 8, 1, 6. This quadruple is 99, 16, 12, 101:

$$
\begin{array}{cccc}
12 & 4 & 3 & 13 \\
99 & 16 & 12 & 101 \; + \\
\hline
1088 & 588 & 441 & 1313
\end{array}
$$

So 1088, 588, 441, 1313 contains an angle which is the sum of the original angle and the angle in the triple with code number 10.

Alternatively we can do the addition using code numbers.

In the last section we saw how to add code numbers instead of the quadruples themselves.

In the present example we obtained d = ½, giving 8, 1, 6 as the code numbers of the quadruple to be added.

If we now add the code numbers:

$$
\begin{array}{cccc}
 & & 1 & \\
4 & & & 3 \; + \\
 & \frac{1}{2} & & \\
\hline
98 & 37\frac{1}{2} & 73\frac{1}{2} & = \; 196, 75, 147
\end{array}
$$

or, alternatively:

$$
\begin{array}{cccc}
 & & 2 & \\
8 & & & 6 \; + \\
 & 1 & & \\
\hline
98\times8 & 100\times3 & 98\times6 & = \; 196, 75, 147.
\end{array}
$$

And the code numbers of 1088, 588, 441, 1313 are 196, 75, 147.

EXAMPLE 8 Given that the Sun was at the position given by the quadruple whose codenumbers are 23,5,10 (which it was on 1978 April 13th), find its rectangular coordinates when it has moved through 43°36' (which was on 1978 May 28th).

The Sun's "orbit" is inclined at about 23°27' to the reference (equatorial) plane. This orbit is called the ecliptic. All quadruples generated from the code numbers 23,d,10 are in the ecliptic. We consider in the next section the problem of finding the values of c and e for a particular given orbit inclination and we also derive the code numbers 23,d,10 for the inclination of the ecliptic.

Now we require the code number of 43°36'. In Chapter 4 we showed how to find the code number of a given angle and so we simply state here that it is k=2.5 (see Section 4.4).

Next we find the value of d: $d = \frac{\sqrt{23^2+10^2}}{2.5} \approx 10$.

So we add the code numbers 23,5,10 and 23,10,10:

$$
\begin{array}{ccc}
 & 5 & \\
23 & & 10+ \\
 & 10 & \\
\hline
579\times23 & 629\times15 & 5790
\end{array} = \underline{4439, 3145, 1930} \text{ which}
$$

are the code numbers of the final position.

Finally, if required, we can generate the quadruple from these code numbers. This is not difficult using the Vedic formula *Vertically and Crosswise,* and working from left to right (see Reference 8, Pages 27, 41, 68) we can easily obtain the final coordinates to 4 figures; these are 1354, 2791, 1214.

The angle between this direction and the correct one is 18'. We have of course rounded off in three places as well as using the approximate code numbers 23,d,10, but the method is exact and so a more accurate answer is obtainable by keeping more significant figures.

7.5   TO OBTAIN A QUADRUPLE WITH A GIVEN INCLINATION

In the example above we used the code numbers 23, d, 10 to represent all quadruples in the plane of the ecliptic, relative to a rectangular, equatorial reference system. These numbers once obtained can be used whenever calculations involving this plane are to be made. Standard planes* whose code numbers would be worth listing would include:

1. inclination of the ecliptic to the equator,
2. inclination of the equator to a particular horizon,
3. inclination of the various planetary orbits to the ecliptic.

We need a way of obtaining these code numbers to any desired degree of accuracy.

In fact the inclination described by the code numbers 23,d,10 is $\tan^{-1}(\frac{10}{23})$ which is 23°30' and not 23°27'. The following method can be used to obtain the code numbers for any inclination to any desired degree of accuracy.

We will take the inclination 23°27' as our example and generate code numbers for this to various degrees of accuracy. It will be seen from the geometry of the problem that we want two integers whose ratio is tan23°27' (see Page 106). We therefore begin by expressing tan 23°27' as a continued fraction. We get:

$$\tan 23°27' = \frac{1}{2+} \ \frac{1}{3+} \ \frac{1}{3+} \ \frac{1}{1+} \ \frac{1}{1+} \ \frac{1}{1+} \ \frac{1}{3+}.$$

This gives the convergents: $\frac{1}{2} \quad \frac{3}{7} \quad \frac{10}{23} \quad \frac{13}{30} \quad \frac{23}{53} \quad \frac{36}{83} \quad \frac{131}{302}.$

---

* There are small or slow variations in the inclinations in some cases.

These fractions give the desired ratio to increasing degrees of accuracy. The third of these, $\frac{10}{23}$, is the ratio used in Example 8. Had we used the sixth, $\frac{36}{83}$, we would have been working with an angle of 23°27'. So any quadruple generated by 83, d, 36 will be in an orbit inclined at 23°27' (to the nearest minute of arc).

A Note on Continued Fractions

Simple examples of continued fractions are: $\frac{2}{3+\frac{4}{5}}$ and $\frac{1}{3+\frac{1}{5+\frac{1}{7}}}$. There can be any number of

fractions within the continued fraction and they can in fact in infinite in length. Continued fractions can be written in a more convenient form. The above two examples could be written as $\frac{2}{3+}\,\frac{4}{5}$ and $\frac{1}{3+}\,\frac{1}{5+}\,\frac{1}{7}$. We will deal only with continued fractions in which all the numerators are 1.

Continued fractions can be very useful. For example, it may sometimes happen that we have got a code number pair representing some angle, but the numbers are too large and we would like to find another pair with smaller numbers which also represent the same angle, though they may be a little less accurate. Suppose, for example, that we have the code numbers 267,101 and we want smaller numbers with a similar ratio. We can convert 267,101 into a continued fraction, obtain the successive convergents to it, as shown below, and select the one we want.

We give, without proof, the following methods of converting to a continued fraction, and of obtaining the successive convergents.

$$(65)\quad(36)\quad(29)\quad(7)\quad(1)\quad(0)$$

$$\frac{101}{267}=\frac{1}{2+}\quad\frac{1}{1+}\quad\frac{1}{1+}\quad\frac{1}{1+}\quad\frac{1}{4+}\quad\frac{1}{7}$$

The numerators in this continued fraction are all 1's, we only have to find the denominators. We begin by dividing 267 by 101: since it divides 2 times we put 2 as the first denominator, and the remainder, 65, in placed in brackets as shown. Then $101\div65 = 1$ remainder 36, so 1 and 36 give the denominator and the bracketed quantity in the next column. Continuing in this way we obtain the complete continued fraction, or we can stop whenever we like.

Next we would like to obtain the successive convergents of this continued fraction. These are the first column of the continued fraction, the first two columns expressed as a single fraction, the first three columns, and so on.

$$\frac{101}{267}=\frac{1}{2+}\quad\frac{1}{1+}\quad\frac{1}{1+}\quad\frac{1}{1+}\quad\frac{1}{4+}\quad\frac{1}{7}$$

$$\frac{1}{2}\quad\frac{1}{3}\quad\frac{2}{5}\quad\frac{3}{8}\quad\frac{14}{37}\quad\frac{101}{267}\quad\text{successive convergents}$$

We can see that the first two convergents are $\frac{1}{2}$ and $\frac{1}{3}$, and these are placed under the first two columns as shown. Each subsequent convergent can be obtained from the denominator above it and the two previous convergents. The numerator of the third convergent is obtained by multiplying the denominator above it by the previous convergent's numerator and adding the numerator before that. We get $1 \times 1 + 1 = 2$ for the third convergent's numerator.

The denominator of the third convergent is similarly obtained by multiplying the denominator above it by the previous convergent's denominator and adding the denominator before that. We get $1 \times 3 + 2 = 5$ for the third convergent's denominator.

For the fourth convergent we get: $1 \times 2 + 1 = 3$ (numerator), and $1 \times 5 + 3 = 8$ (denominator).

For the fifth convergent: $4 \times 3 + 2 = 14$, $4 \times 8 + 5 = 37$.

The sixth convergent gives the initial fraction: $7 \times 14 + 3 = 101$, $7 \times 37 + 8 = 267$.

Thus, all these convergents approximate $\frac{101}{267}$ to different degrees and $\frac{14}{37}$ is the closest distinct fraction to $\frac{101}{267}$.

The accuracy lost in making this approximation will be given by $\frac{101}{267} - \frac{14}{37}$ and since the determinant formed by any two successive convergents is equal to 1 or -1 (that is $\begin{vmatrix} 14 & 101 \\ 37 & 267 \end{vmatrix} = 1$) the accuracy lost is given by $\frac{1}{37 \times 267} = 0.0001$.

7.6   ANGLE BETWEEN TWO DIRECTIONS

Since two quadruples define two directions in 3-dimensional space there will be an angle between these two directions.

So given 8,4,1,9 and 6,3,2,7 we find the angle between them as follows.

We subtract the quadruples using formula (1) from Page 100. But, as stated on that page we combine the three middle elements together to obtain a triple. That is the quintuple a,b,c,d,e becomes the triple a, $\sqrt{b^2 + c^2 + d^2}$,e.

8	4	1	9
6	3	2	7 −

$$A\hat{O}B: \quad 8 \times 4 + 4 \times 3 + 1 \times 2, \quad \sqrt{(4 \times 2 - 1 \times 3)^2 + (1 \times 6 - 8 \times 2)^2 + (8 \times 3 - 4 \times 6)^2} \quad 9 \times 7$$

$$= \quad 62 \quad\quad\quad \sqrt{125} \quad\quad\quad 63$$

We combine the 3 middle elements by squaring them, adding and taking the square root.

So the triple 62, $\sqrt{125}$, 63 represents angle AOB (see proof below).

Note that the middle element of the triple can be calculated more easily from the other elements than by the method shown above.

That is $63^2 - 62^2 = 125$, or $(63 + 62)(63 - 62) = 125$.

That this gives the required triple can be shown as follows.

We require a triple for $\overset{\wedge}{A O B}$.

Consider the xy-plane:

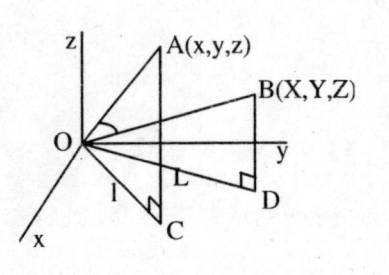
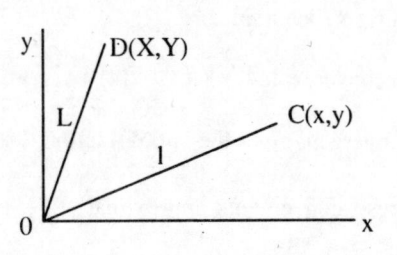

Rotate A, B, C and D clockwise about the z-axis by angle x, y, 1 to A', B', C', D'.

∴ for D':

D :	X	Y	L
C :	x	y	1
D' :	$\frac{xX+yY}{1}$	$\frac{Yx-Xy}{1}$	L

Then we have:

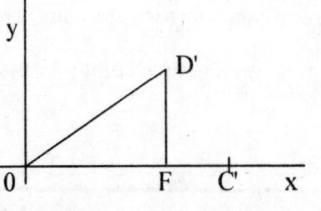

Consider now the xz-plane: drop a perpendicular from B' to meet the xz-plane at B'' and let OB''=p. Then for $\overset{\wedge}{A'O B''}$:

A' :	1	z	r
B'' :	(xX+yY)/l	Z	p
$\overset{\wedge}{A'OB''}$:	$xX + yY + zZ$,	-,	rp

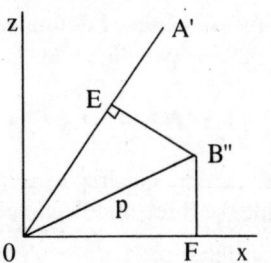

∴ $OE = \frac{xX+yY+zZ}{r}$ and $OB' = R$ ∴ $\overset{\wedge}{A'OB'}$) $\frac{xX+yY+zZ}{r}$, -, $R = \overset{\wedge}{AOB}$

∴ $\overset{\wedge}{AOB}$) xX+yY+zZ, -, rR.

Then the third element of this triple is given by:
$(EB')^2 = (x^2+y^2+z^2)(X^2+Y^2+Z^2) - (xX+yY+zZ)^2$
which becomes $(EB')^2 = (xY - yX)^2 + (xZ - zX)^2 + (yZ - zY)^2$.

∴ $\overset{\wedge}{AOB}$) (xX+yY+zZ), $\sqrt{(xY - yX)^2 + (xZ - zX)^2 + (yZ - zY)^2}$, rR

Spherical Triangles

Three quadruples define three directions in space, which define a spherical triangle on, say, the celestial sphere. If we want a triple for any side of that triangle we can use the method described above. If we want a triple for any angle in that triangle we can proceed as indicated in the note on Page 73, or alternatively as follows.

EXAMPLE 9

Find the angle B in the spherical triangle defined by the quadruples:

$$
\begin{array}{c|cccc}
A' & 1 & 1 & 2 & -- \\
B' & 6 & 3 & 2 & 7 \\
C' & 11 & 4 & 3 & --
\end{array}
$$

For the first element of B we multiply in the first and third rows, the first and second rows and the second and third rows as shown below:

$$(1.11 + 1.4 + 2.3).7^2 - (1.6 + 1.3 + 2.2)(6.11 + 3.4 + 2.3) = -63.$$

For the second element we multiply the magnitude of the determinant of the 3 by 3 array by 7, the last element of the quadruple B:

$$|1.(3.3 - 2.4) - 1.(6.3 - 2.11) + 2.(6.4 - 3.11)|.7 = 91.$$

Dividing out the common factor 7 we get B) –9, 13, -.

INCLINATION OF PLANETARY ORBITS

The first point of Aries, ♈, always lies in the plane of the ecliptic so that if it is not in the plane of the planet's orbit it is because the orbits of the Earth and the planet are inclined to each other. The orbit of Venus is inclined at about 3° to the earth's orbit.

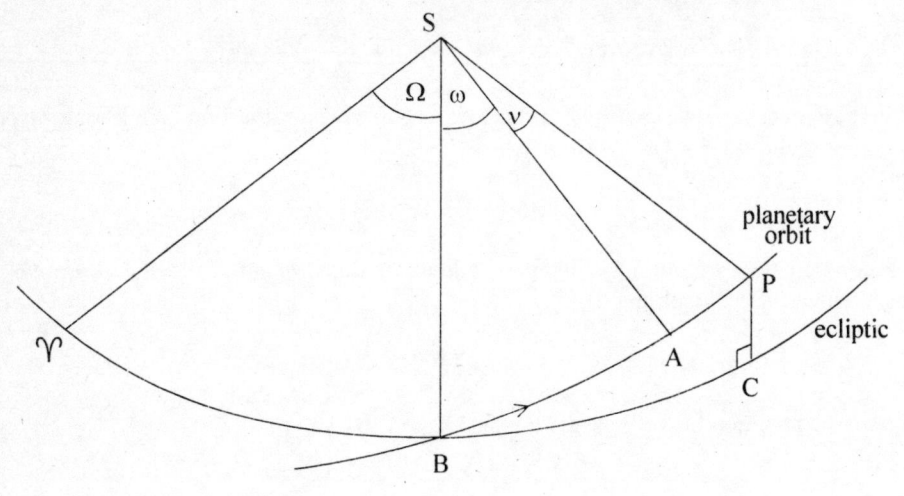

Figure 10: Inclination of a planetary orbit to the ecliptic

In Figure 10, S represents the position of the Sun, ♈BC the ecliptic and BAP the orbit of a planet, say Venus.

i, the inclination of the orbits, is the angle between the tangents to the two arcs at B i.e. for Venus i ≈ 3°.

B is the node of the planet's orbit, i.e. where the planet crosses the ecliptic when going from south to north of it.

A is the perihelion point of the planet, so $v = A\hat{S}P$.

$\Omega = ♈\hat{S}B$ is the longitude of the planet's node, and $\varpi = \Omega + \omega$, where $\omega = B\hat{S}A$ is called the argument of perihelion. ϖ is therefore the longitude of the planet's perihelion, but measured along the ecliptic from ♈ to B, and then along the planet's orbit to A.

Now in Example 4 on page 58 we found $L = \varpi + v = \Omega + \omega + v$. But Ω is not in the same plane as ω and v. The ecliptic longitude, l, is ΥSC, i.e. it is measured along the ecliptic, and this is the quantity listed for planetary longitudes in the Ephemerides.

To find l we find $L - \Omega$ (i.e. $\omega + v$) and find BC in the spherical triangle PBC:

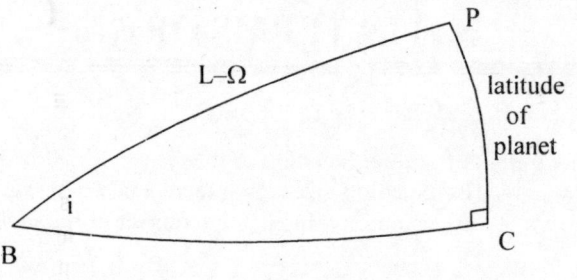

Then $l = \Omega + BC$, Ω and i are listed for each of the planets below.
Using a formula from spherical trigonometry:

$$\tan BC = \tan(L - \Omega)\cos i$$

Therefore, for Example 4 we have: $\tan BC = \tan(220.394 - 76.545)\cos 3.394$.
From which $BC = 143.897°$.

Therefore $l = \Omega + BC = 76.545 + 143.897 = 220.442 = \mathbf{220°27'}$.

The latitude of the planet is also available from this triangle:
$\sin PC = \sin i \sin(L - \Omega)$.

Values of Ω and I

	Ω	i
MERCURY	48.153	7.004
VENUS	76.545	3.394
MARS	49.442	1.850
JUPITER	100.297	1.304
SATURN	113.526	2.489
URANUS	73.924	0.773
NEPTUNE	131.615	1.771
PLUTO	110.216	17.132

DERIVATION OF FORMULA FOR $c(v)$ IN TERMS OF c(M)

We will obtain this formula from the Equation of the Centre which is derived in most books on Spherical Astronomy. This equation gives the true anomaly in terms of the mean anomaly and ascending powers of the eccentricity. Including terms up to and including e^2 it is:

$$v = M + 2e\sin M + \tfrac{5}{4}e^2\sin 2M.$$

$$\therefore \tan\tfrac{v}{2} = \tan\left(\tfrac{M}{2} + e\sin M + \tfrac{5}{8}e^2\sin 2M\right).$$

Then using the trigonometric identity $\tan(A+B+C) \equiv \dfrac{\tan A+\tan B+\tan C-\tan A\tan B\tan C}{1-\tan A\tan B-\tan A\tan C-\tan B\tan C}$

and expanding the tan functions as series, but neglecting terms in e^3 or higher powers of e we get:

$$\tan\tfrac{v}{2} = \frac{\tan\left(\tfrac{1}{2}M\right)+e\sin M+\tfrac{5}{8}e^2\sin 2M}{1-\tan\left(\tfrac{1}{2}M\right)e\sin M-\tan\left(\tfrac{1}{2}M\right)\tfrac{5}{8}e^2\sin 2M}$$

Then putting c = c(M) = cot½M, and since $\sin M = \dfrac{2c}{c^2+1}$ and $\cos M = \dfrac{c^2-1}{c^2+1}$ (see Page 48, diagram), we have:

$$\tan\tfrac{v}{2} = \frac{\tfrac{1}{c} + \tfrac{2ce}{c^2+1} + \tfrac{5e^2}{4}\times\tfrac{2c(c^2-1)}{(c^2+1)^2}}{1-\tfrac{1}{c}\times\tfrac{2ce}{c^2+1} - \tfrac{1}{c}\times\tfrac{5e^2}{4}\times\tfrac{2c(c^2-1)}{(c^2+1)^2}}\ .$$

$$\therefore c(v) = c\left(\frac{1-\tfrac{2e}{c^2+1}-\tfrac{5c(c^2-1)}{2(c^2+1)^2}}{1+\tfrac{2ce}{c^2+1}+\tfrac{5e^2c^2(c^2-1)}{2(c^2+1)^2}}\right).$$

Finally, by dividing the denominator of this fraction into the numerator we arrive at:

$$c(v) = c\left(1-2e+\frac{e^2(3c^2+5)}{2(c^2+1)}\right), \text{ which may be written as}$$

$$c(v) = c\left(1-2e+e^2\left(1.5+\tfrac{1}{c^2+1}\right)\right).. \qquad \qquad \dots (1)$$

Similarly, if terms up to and including e^3 are retained we may obtain:

$$c(v) = c\left(1 - 2e + e^2\left(1.5 + \tfrac{1}{c^2+1}\right) - e^3\left(1 - \tfrac{\frac{2}{3}c^2-2}{(c^2+1)^2}\right)\right).$$

Aphelion

These formulae for $c(v)$ use perihelion as base. If aphelion is to be the base we may obtain corresponding formulae by considering the relation between $c(v)$ measured from perihelion and $c(v)$ measured from aphelion, and also between $c(M)$ measured from perihelion and $c(M)$ measured from aphelion. But:

v (perihelion) $= v$(aphelion) $+ \pi$, and
M(perihelion) $= M$(aphelion) $+ \pi$.

Therefore, as shown on page 121:

$$c(v \text{ measured from aphelion}) = \frac{-1}{c(v \text{ measured from perihelion})} \text{ and similarly for M.}$$

If we make these two substitutions in equation (1) on the previous page we obtain:

$$c(v) = \frac{c}{\left(1 - 2e + e^2\left(1.5 + \tfrac{c^2}{c^2+1}\right)\right)}.$$

where $c = c(M)$ and both $c(v)$ and $c(M)$ are taken as measured from aphelion.

On division this gives:

$$c(v) = c\left(1 + 2e + e^2\left(1.5 + \tfrac{1}{c^2+1}\right)\right). \qquad \ldots (2)$$

APPENDIX III

CALCULATION OF THE RADIUS VECTOR

On page 63 we found the geocentric longitude of Saturn on 1985 Feb 14 to be 238°2'.
To obtain a more accurate answer we should use the actual distances (the radius vectors) of
the Earth and Saturn from the Sun, rather than the mean values.

The distance of a body in an elliptical orbit from its primary is given by $r = \dfrac{a\left(1-e^2\right)}{1+e\cos v}$ where r,
a, e, v, have the same meanings as previously, except that v must be measured from the
perihelion point. This formula is derived in most books on positional astronomy.

Now since $\cos v = \frac{c^2-1}{c^2+1}$ where $c = c(v)$, $r = \dfrac{a\left(1-e^2\right)}{1+e\frac{c^2-1}{c^2+1}} = \dfrac{a\left(1-e^2\right)}{1+e-\frac{2e}{c^2+1}}$.

So by substituting the code numbers of the true anomaly of the Earth and Saturn we can
obtain their distances from the Sun at the required time:

For Saturn $c(v) = 0.3620$, see Page 57.
so that r = 9.934.

And for the Earth $c(v) = 2.615$, see Page 56
so that r' = 0.9874.

Therefore $R = \frac{r}{r'} = 10.061$.

Therefore $c(2\theta) = \frac{10.061\times613-35}{612} = 10.020$.

Then $A(10.02) = 0.1993 + \dfrac{2(10-10.02)}{10\times10.02+1} = 0.1989 = 2\theta$.

So $\theta = 5°42'$.

And $\lambda = 232°1' + 5°42' = \textbf{237°43'}$.

This is still 1' short of the correct value, but this due to rounding errors since we found L, L'
etc. to the nearest minute. Better results can be obtained by calculating to one more
significant figure.

We now consider Example 8, Page 64 in which the geocentric longitude of Mars on 1985 Dec 1 was found. We first find r, r' at this time:

For Mars we found $p(v) = 5.755$ (Page 59), but this was with v measured from aphelion. The true anomaly measured from perihelion would be the true anomaly from aphelion plus π. The code number would then be $\frac{-1}{5.755} = -0.1738$. This is demonstrated as follows.

If $c^2 - 1$, $2c$, $c^2 + 1$ is a triple with code number c,
then $1 - c^2$, $-2c$, $c^2 + 1$ is a triple whose angle has been increased by π radians.

And the code number of this is 2, $-2c = \frac{-1}{c}$.

Then, $r = 1.6560$.

And for the Earth $c(v) = -3.2014$ (Page 57), so that $r' = 0.9862$.

Therefore, $R = 1.6792$.

We should also obtain better values for x,y,z.

We had $c(\alpha) = c(1.8637) = c(\frac{\pi}{2} + 0.2929)$.
Now $0.2929 - 0.2838 = 0.0091$ and $\frac{2}{0.0091} = 220$.

Therefore, $c(0.2929) = 7 + 220 = 30\frac{1}{2}, 4\frac{1}{2} = 61,9$.

Therefore, $c(\alpha) = 1 + 61,9 = 26,35$.

And $x = -549$, $y = 1820$, $z = 1901$

Therefore, $c(2\theta) = \frac{1.6792 \times 1901 - -549}{1820} = 2.0556$.

And $A(2.0556) = 0.9273 + \frac{2(2-2.056)}{2 \times 2.056 + 1} = 0.9054$ radians.

Therefore, $\theta = 25°56'$.
Therefore, $\lambda = 175°31' + 25°56' = \textbf{201°27'}$.

PROOFS OF SPHERICAL TRIANGLE FORMULAE (CHAPTER 6)

For the purpose of these proofs we will denote the triples for the sides and angles of a spherical triangle as follows:

a	l	m	n		A	L	M	N
b	p	q	r		B	P	Q	R
c	s	t	u		C	S	T	U

THE COSINE RULE

Suppose we have two sectors BOC and BOA' containing angles a and c respectively that lie in the x-y plane.

Let the point A' be represented by the triple: A')es, et, eu
and C by: C)fl, fm, fn.

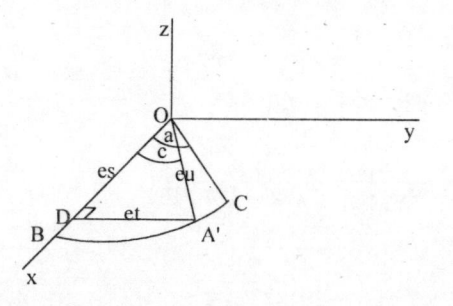

Consider now a plane perpendicular to the x-axis, through D (diagram above).
In this plane, with D as origin, A' is given by: A')et, 0, et.

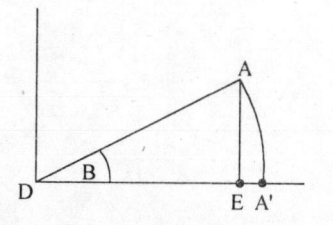

We now rotate A' in this plane, about the x-axis by an angle: B) P, Q, R., to the point A (see diagram above).

Then A' | et   0   et

      B | P   Q   R   +

      A | $\frac{Pet}{R}$ $\frac{Qet}{R}$ et

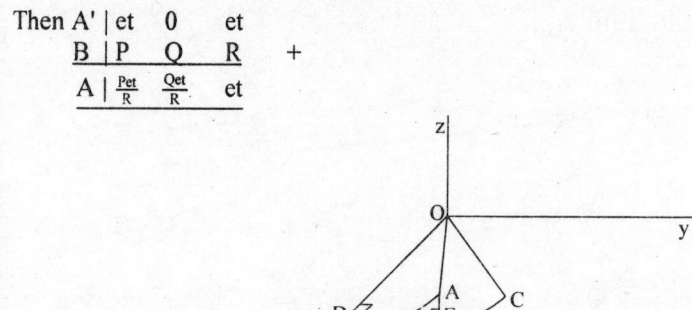

Now let E be the foot of the perpendicular from A onto the x-y plane.

Then since A has coordinates $\left(es, \frac{Pet}{R}, \frac{Qet}{R}, eu\right)$, by using the formula for the angle between

two lines (see page 113) $A\hat{O}C$ is given by:

A | es $\frac{Pet}{R}$ $\frac{Qet}{R}$ eu

C | fl   fm   0   fn

$A\hat{O}C$ | esfl + $\frac{Petfm}{R}$,   -,   eufn   = Rls + Pmt, - , nRu

Therefore in the spherical triangle ABC, b)Rls+Pmt, - , nRu.

Or, diagrammatically:

a| l   m   n

B| P   Q   R $\mapsto$ = b) p - r

c| s   t   u

Thus since b)p,q,r, $\frac{P}{r} = \frac{Pls+Pmt}{nRu}$.

Therefore, $\frac{P}{r} = \frac{ls}{nu} + \frac{mt}{nu} \times \frac{P}{R}$.

Therefore, $\frac{P}{R} = \frac{pnu-rls}{rmu} \times \frac{nu}{mt} = \frac{pnu-rls}{rmt}$.

That is:

a| l   m   n

b| p   q   r $\mapsto$ = B) P - R

c| s   t   u

THE SINE RULE

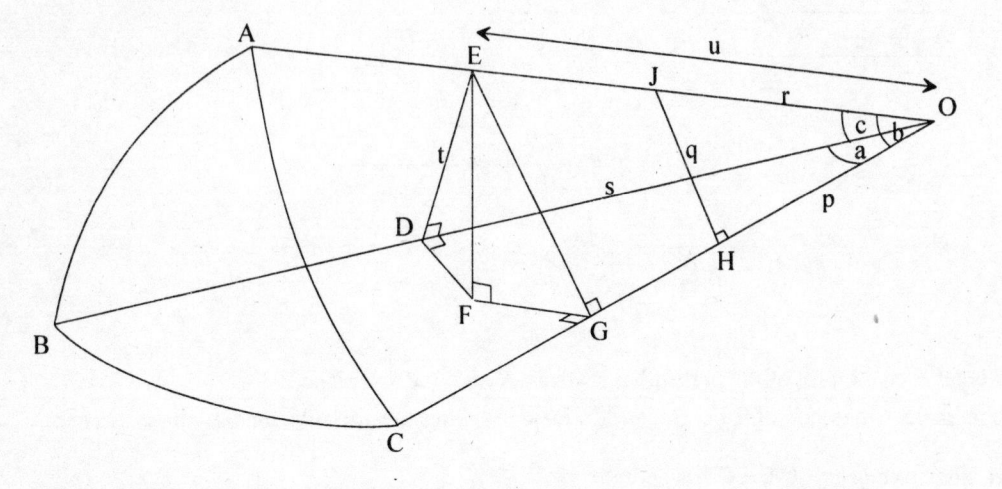

Let the right-angled triangles ODE and OHJ have dimensions given by the triples:
c)s,t,u and b)p,q,r.

Draw EG parallel to JH and drop the perpendicular EF onto the plane OBC.
Join FD and FG.

Then, $EG = \frac{qu}{r}$.

Also $O\hat{G}F = 90°$ since EG and EF are perpendicular to OG, so that the plane EFG and therefore the line FG are at right angles to OG.

Similarly $O\hat{D}F = 90°$.

But B)P,Q,R = $F\hat{D}E$ and C)S,T,U = $F\hat{G}E$, therefore, equating the two expressions for EF obtained from each triangle:

$$\frac{Qt}{R} = \frac{T}{U} \times \frac{qu}{r},$$

and therefore $\frac{Q}{R} = \frac{quT}{rtU}$, or B) - , quT, rtU.

Diagrammatically:

$$
\begin{array}{c|ccc}
b & p & q & r \\
c & s & t & u \\
C & S & T & U
\end{array}
\quad \mapsto \quad
\text{= B)- Q R}
$$

Also $\frac{q}{r} = \frac{QUt}{RTu}$, therefore b) - , QUt, Rtu

$$\begin{array}{c|ccc} B & P & Q & R \\ C & S & T & U \\ c & s & t & u \end{array} \longmapsto$$ $= b) - q \quad r$

THE COTANGENT RULE

This can be derived from the Cosine and Sine Rules.

From the Cosine Rule for b and c we get: $\dfrac{P}{r} = \dfrac{Pmt+Rls}{nRu}$ (1)

$$\dfrac{s}{u} = \dfrac{Sqm+pUl}{rUm}$$ (2)

And from the Sine Rule we have: $\dfrac{t}{u} = \dfrac{TRq}{UQr}$ (3)

Substituting for $\dfrac{s}{u}$ and $\dfrac{t}{u}$ from (2) and (3) into (1):

$$\dfrac{P}{r} = \dfrac{Pm}{nR} \times \dfrac{TRq}{UQr} + \dfrac{1}{n} \times \dfrac{Sqm+pUl}{rUn} \text{ therefore}$$

$$\dfrac{P}{Q} = \left(\dfrac{p}{r} - \dfrac{lSqm+pUl^2}{n^2rU} \right) \times \dfrac{nUr}{mTq} = \dfrac{pn^2U - lSqm - pUl^2}{nmTq} = \dfrac{pUm^2 - lSqm}{nmTq}$$

as $n^2 - l^2 = m^2$.

$$\therefore \dfrac{P}{Q} = \dfrac{pmU-qlS}{qnT} \text{ and } \dfrac{p}{q} = \dfrac{PTn+QSl}{QUm}.$$

That is:

$$\begin{array}{c|ccc} b & p & q & r \\ a & l & m & n \\ C & S & T & U \end{array} \longmapsto$$ $= B) P \quad Q \ -$

And

$$\begin{array}{c|ccc} B & P & Q & R \\ C & S & T & U \\ a & l & m & n \end{array} \longmapsto$$ $= b) p \quad q \ -$

RIGHT-ANGLED TRIANGLES

If in a spherical triangle $\hat{C} = 90°$, then since C is given by the triple S,T,U, S=0 and T=U=1.

We can obtain eight of the ten formulae for right-angled triangles from the formulae for general spherical triangles.

For example, from the Cosine formula to find c:

$$
\begin{array}{lll}
b \mid p & q & r \\
C \mid 0 & 1 & 1 \\
a \mid l & m & n
\end{array}
\qquad c \mid pl, \quad -, \quad rn \quad \text{or} \quad \frac{s}{u} = \frac{pl}{rn} \qquad (1)
$$

Similarly, from the Sine Rule: $\quad \frac{Q}{R} = \frac{qu}{rt}$ (8)

$$\frac{M}{N} = \frac{um}{tn} \qquad (7)$$

From the Cotangent Rule: $\qquad \frac{P}{Q} = \frac{pm}{qn}$ (4)

$$\frac{L}{N} = \frac{sq}{tp} \qquad (9)$$

From the Polar Cosine Rule: $\qquad \frac{P}{r} = \frac{PN}{RM}$ (5)

$$\frac{l}{n} = \frac{LR}{NQ} \qquad (6)$$

$$\frac{s}{u} = \frac{PL}{QM} \qquad (2)$$

The last two formulae may be derived from the above:

from (5) $\frac{P}{R} = \frac{pM}{rN}$, from (1) and (7) $\frac{P}{R} = \frac{sn}{ul} \times \frac{um}{tn} = \frac{ms}{lt}$ (10)

from (2) $\frac{M}{L} = \frac{uP}{sQ}$, from (1) and (4) $\frac{M}{L} = \frac{rn}{pl} \times \frac{pm}{qn} = \frac{mr}{lq}$ (3)

These formulae are shown diagrammatically on Page 83.
The corresponding conventional formulae are:

(1) $\cos c = \cos a \cos b$ (2) $\cos c = \cot A \cot B$

(3) $\tan A = \dfrac{\tan a}{\sin b}$ (4) $\tan B = \dfrac{\tan b}{\sin a}$

(5) $\cos b = \dfrac{\cos B}{\sin A}$ (6) $\cos a = \dfrac{\cos A}{\sin B}$

(7) $\sin A = \dfrac{\sin a}{\sin c}$ (8) $\sin B = \dfrac{\sin b}{\sin c}$

(9) $\cos A = \dfrac{\tan b}{\tan c}$ (10) $\cos B = \dfrac{\tan a}{\tan c}$

PROOF OF CODE-NUMBER PATTERNS

Given the code numbers of a, b, c, A, B, C:

$$
\begin{array}{c|cc}
a & e & f \\
b & g & h \\
c & i & j
\end{array}
\qquad
\begin{array}{c|cc}
A & E & F \\
B & G & H \\
C & I & J
\end{array}
$$

then the triples for a, b, c are:

$$
\begin{array}{c|ccc}
a & e^2 - f^2 & 2ef & e^2 + f^2 \\
b & g^2 - h^2 & 2gh & g^2 + h^2 \\
c & i^2 - j^2 & 2ij & i^2 + j^2
\end{array}
$$

and the triple for A is A) $E^2 - F^2$, $2EF$, $E^2 + F^2$.

Using the pattern given on Page 70 to find A:

$$(g^2 - h^2)(e^2 + f^2)(i^2 + j^2) - (g^2 + h^2)(e^2 - f^2)(i^2 - j^2) = E^2 - F^2.$$

And $2ef(g^2 + h^2)2ij = E^2 + F^2$.

Adding and subtracting these equations:

$$2E^2 = (g^2 - h^2)(e^2 + f^2)(i^2 + j^2) - (g^2 + h^2)(e^2 - f^2)(i^2 - j^2) + 4efij(g^2 + h^2),$$
$$2F^2 = 4efij(g^2 + h^2) - (g^2 - h^2)(e^2 + f^2)(i^2 + j^2) + (g^2 + h^2)(e^2 - f^2)(i^2 - j^2).$$

After expanding, cancelling and factorising we get:

$$E^2 = \{e(gj + hi)\}^2 - \{f(gi - hj)\}^2,$$
$$F^2 = \{f(gi + hj)\}^2 - \{e(gj - hi)\}^2.$$

The proofs of the patterns used in Examples 9, 10, 11 follow very similar lines:

we write out the triples involved in terms of their code numbers, apply the appropriate formula from Section 6.1 (for Examples 8, 9 we use the Cosine Rule, and for Examples 10, 11 the Polar Cosine Rule), add and subtract the two equations thus obtained, expand, cancel and factorise.

The pattern used in Examples 12, 13 is derived as follows.

For a, b, A the triples are:
$$
\begin{array}{c|ccc}
a & e^2 - f^2 & 2ef & e^2 + f^2 \\
b & g^2 - h^2 & 2gh & g^2 + h^2 \\
A & E^2 - F^2 & 2EF & E^2 + F^2
\end{array}
$$

The triple for B is B) $G^2 - H^2$, $2GH$, $G^2 + H^2$ and the triple complementary to B is B(CT)) $2GH$, $G^2 - H^2$, $G^2 + H^2$.

Let the code numbers of B(CT) be G', H'.

Then applying the formula given on Page 72 (Sine Rule to find an angle):

$2ef(g^2 + h^2)2EF = G'^2 = H'^2$ and
$(e^2 + f^2)2gh(E^2 + F^2) = G'^2 + H'^2$

Then adding and subtracting etc. we get:

$G'^2 = (egE + fhF)(fgE + ehF) + (ehE + fgF)(fhE + egF),$
$H'^2 = (egE - fhF)(fgE - ehF) + (ehE - fgF)(fhE - egF).$

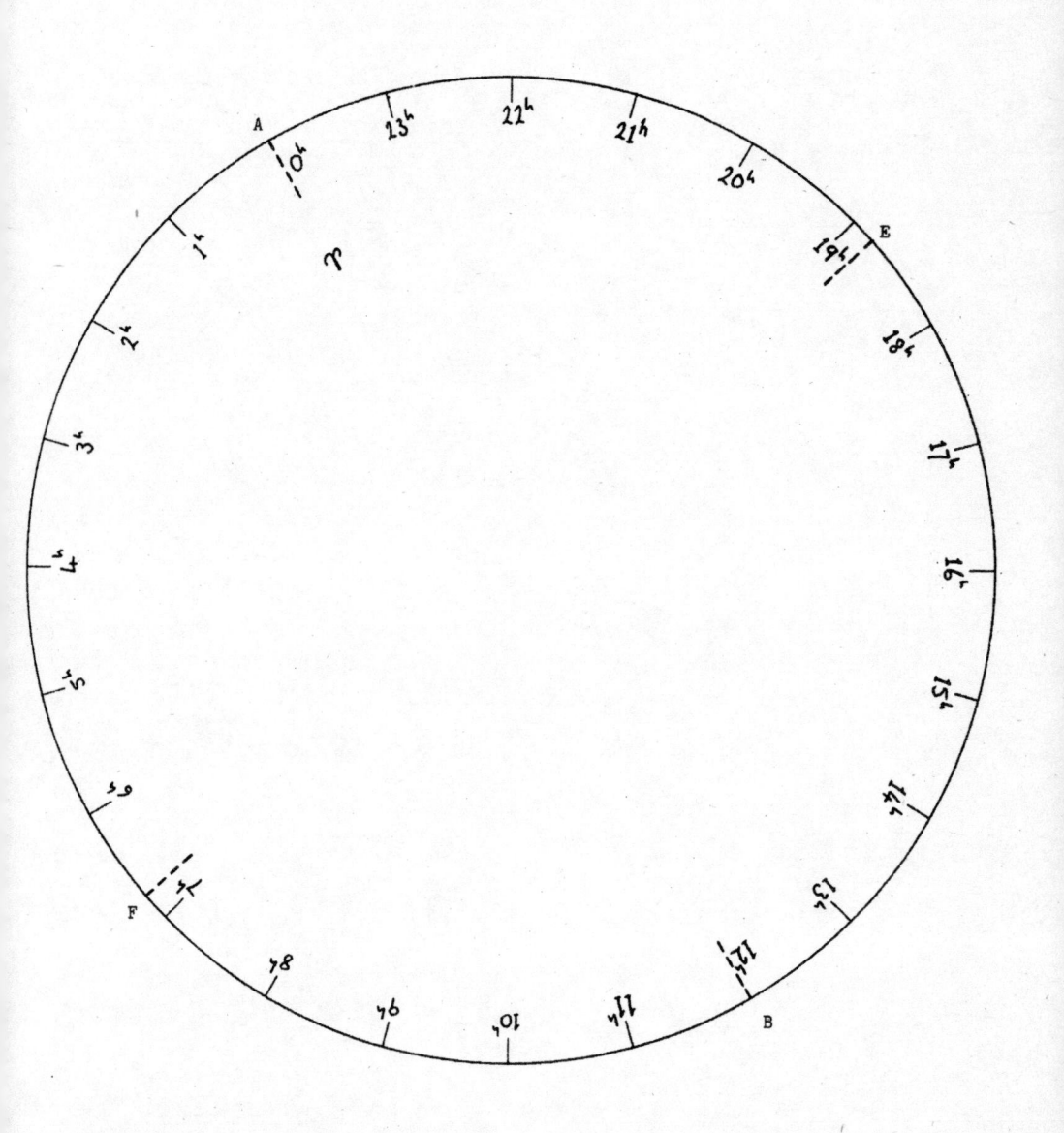

The Planet Finder – Circle 1
The plane of the equator divided into 24 hours beginning at ♈
Each hour is equivalent to 15°

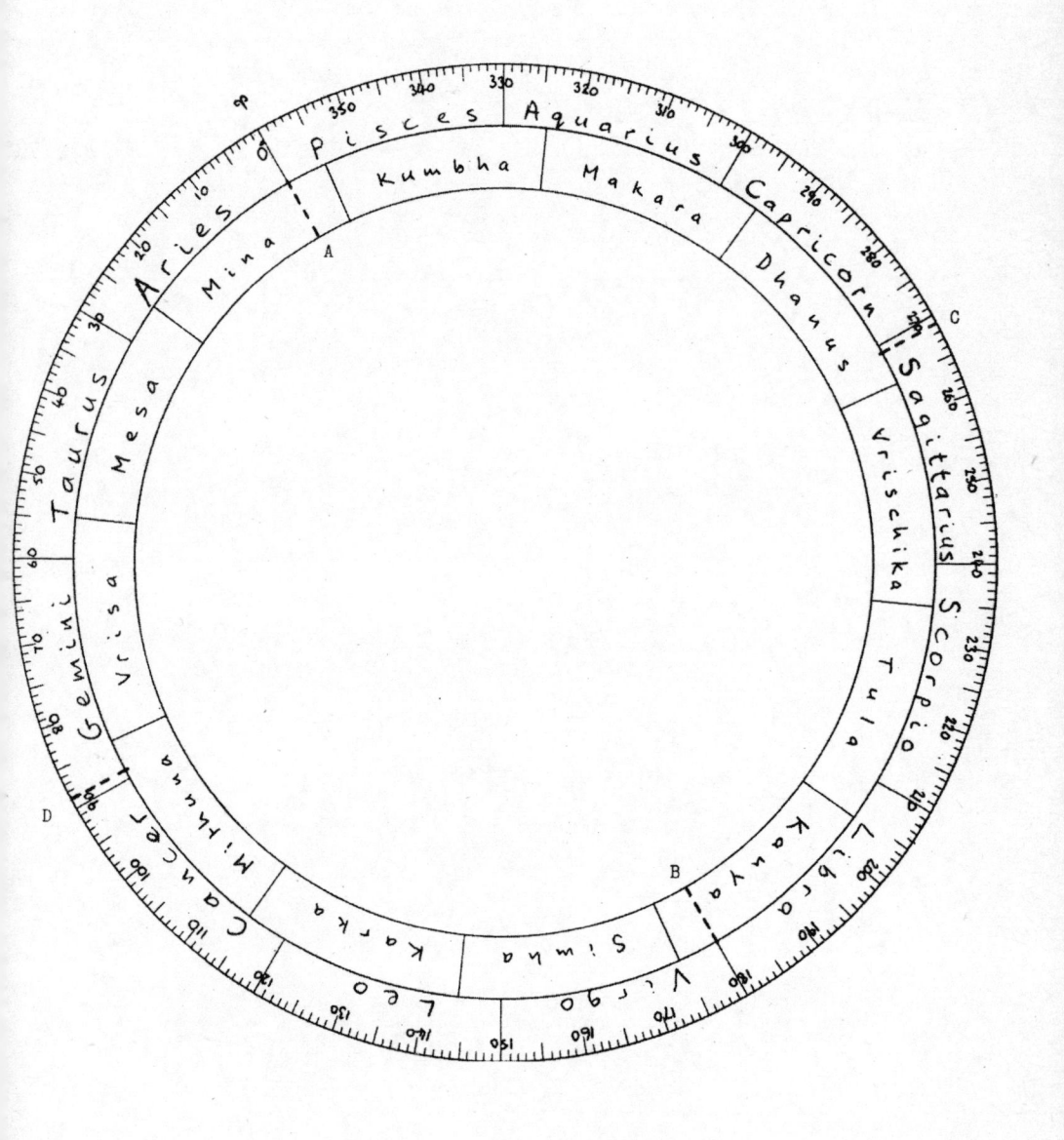

The Planet Finder – Circle 2

The plane of the ecliptic divided into 360° starting at ♈. The outer circle shows the 12 western zodiacal signs. The inner circle shows the 12 Indian signs (Aries = Mesa etc.). Unlike the western signs which move relative to the stars the Indian signs are fixed on the celestial sphere. The difference the two systems is about 24½° at present.

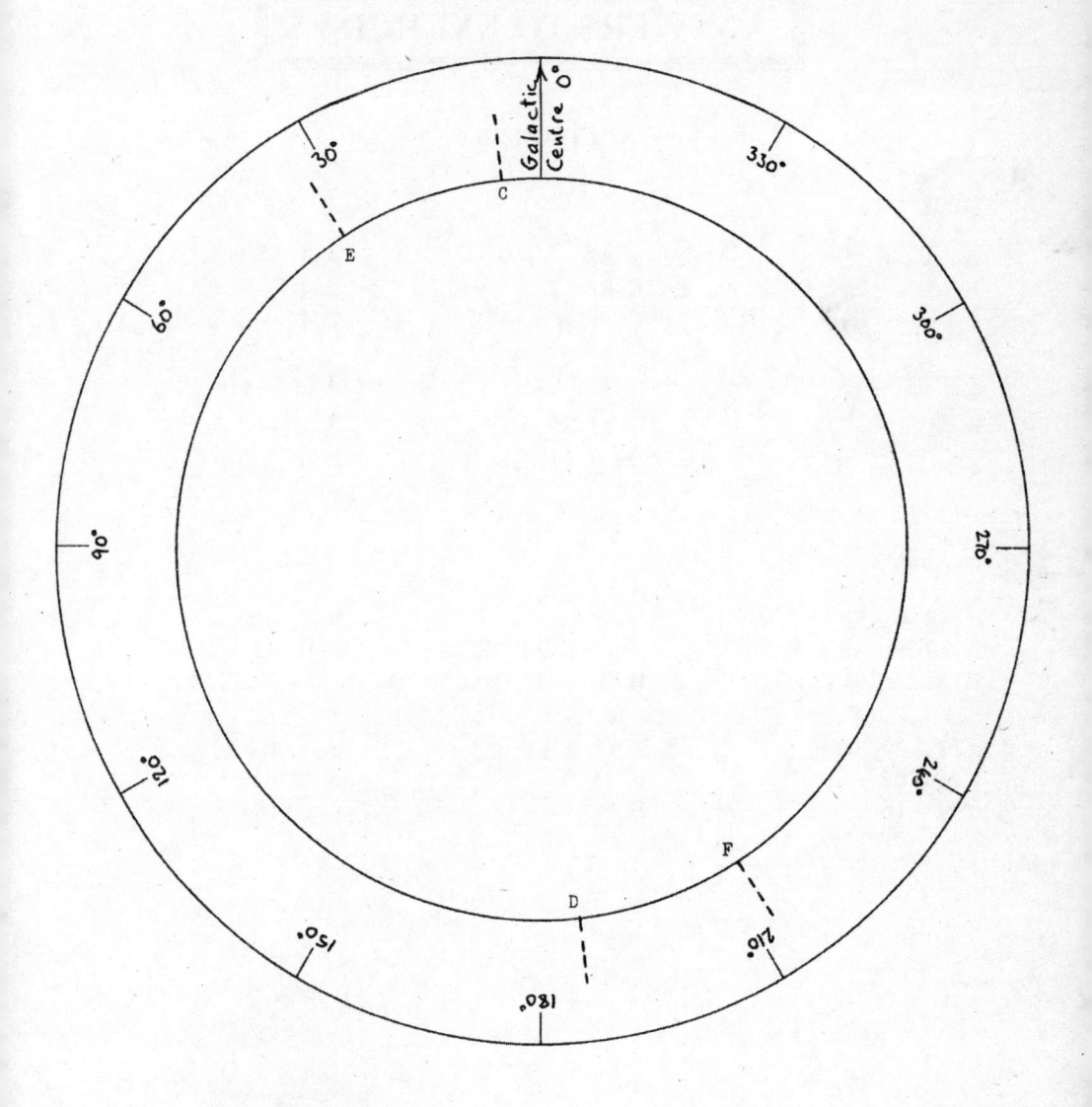

The Planet Finder – Circle 3
The Galactic plane divided into 360° measured from the galactic center.

CHAPTER 6

Exercise A

1. A 53 -- 85
   B -2 -- 25
   C 10 -- 13

2. a 88 -- 169
   B 61 52 --
   C -1 16 --

3. A 1 -- 3
   B 1 -- 3
   C 1 -- 3

4. a $2-\sqrt{3}$ -- $\sqrt{3}$
   b $\sqrt{3}-2$ -- $\sqrt{3}$
   c $2\sqrt{3}-1$ -- 3

5. a 32 -- 85
   B 25 8 --
   C -20 17 --

6. A 81 -- 119
   B -5 -- 14
   C 9 -- 10

Exercise B

1. c 48 -- 65
   A 20 39 --
   B 36 25 --

2. b 5 -- 8
   A -- 7 20
   B 7 -- 32

3. b 4 1 --
   c 16 13 --
   A 4 -- 13

4. a 13 -- 20
   b 25 -- 52
   c 5 -- 16

5. a -- 3 13
   b 13 9 --
   B 1 3 --

6. a 1 1 $\sqrt{2}$
   b 0 1 1
   c 0 1 1

REFERENCES

Nicholas, Williams, Pickles, *Vertically and Crosswise,* Inspiration Books, 1999.

Tirthaji, B.K., *Vedic Mathematics,* Delhi: Motilal Banarsidass, 1965.

Williams, K.R., *Triples,* Inspiration Books, 1999.

GLOSSARY

Angular advance: See Page 110

Annular eclipse: An eclipse of the Sun when a ring of the Sun's surface is seen around the Moon due to the Moon's angular size being less than that of the Sun.

Anomalistic month: The time passing from when the Moon is closest to the Earth until the next time it is closest to the Earth

Aphelion: The point in a planet's orbit which is furthest from the Sun.

Celestial equator: A circle on the celestial sphere which is a projection of the Earth's equator.

Complementary angles: Two angles that total a right angle (90 degrees or 2 radians).

Conjunction: Two planets are in conjunction when they have the same longitude. They are then seen close together.

Cross-product (CP): See Page 1.

Declination (δ): The angular distance of a heavenly body from the celestial equator.

Direction vector: A vector which is used to define a direction in 2 or 3-dimensional space.

Duplex (D): See Page 9.

Eccentric Anomaly (E): See Page 31 Eccentricity (e) See Page 31.

Eclipse: Interception of the light of a luminous body by the passage of another body.

Ecliptic: The Sun's apparent path in the sky relative to the stars. The plane of the earth's orbit.

Equation of time: The difference between mean solar time as given by a clock and apparent solar time as given by a sundial.

Equatorial coordinates (α, δ): The direction of a heavenly body defined by its Right Ascension and Declination.

Equatorial horizontal parallax (p): The maximum value of parallax possible for a heavenly body from two observations on the surface of a another body.

Equatorial plane: The plane of the Earth's equator.

Equatorial radius: The length of the line from the centre of a rotating body to a point on its equator.

Equinox: The time, or the point, at which the Sun crosses the celestial equator.

First Point of Aries (Υ): The point of intersection of the equator and the ecliptic where the Sun passes from south to north.

Focus: When a body has an elliptical orbit around a primary body the primary body always takes the position called the focus to conform with the inverse square law of gravitational attraction.

Galaxy: A very large collection of stars: our own Galaxy is disc-shaped and contains some hundred thousand million stars.

Geocentric: As viewed from the Earth's centre.

Great circle: A circle on the surface of a sphere which has the greatest diameter.

Heliocentric: As viewed from the Sun's centre.
Hypotenuse: The longest side of a right-angled triangle (opposite the right angle).

Inner planet: A planet whose orbit is within the Earth's orbit.
Integer: The positive and negative whole numbers, including zero.
Iteration: A method which uses successive approximations; a starting value leads to a result which is then used to get the next result and so on.

Local time: The time at a place on the Earth measured from the Sun's transit across the meridian.
Longitude (astronomical) (L): The angular distance of a celestial body from the First Point of Aries measured along the ecliptic.
Longitude (terrestrial) (l): The angle that the meridian through the geographical poles and a point on the Earth's surface makes with the Greenwich (standard) meridian.

Mean Anomaly (M): See Pages 31, 55. Mean motion Average rate of movement.
Mean node: A fictitious node that moves at a constant speed and completes a circuit in the same time as the true node.
Mean solar day: The average value of the interval between two successive returns of the Sun to a meridian.
Meridian: The great circle on the celestial sphere which passes through an observer's zenith and the celestial poles.
Meridian transit: This occurs when a body passes through an observer's meridian.

Newton-Raphson method: An iterative method which often converges very quickly to a required solution.
Node: One of two points where the plane of two orbits appear to cross.
North celestial pole: The point where the line from the Earth's centre to the north pole meets the celestial sphere.

Outer planet: A planet whose orbit is outside the Earth's orbit.

Parallax: See Page 14.
Penumbra: Half shadow: the result of light from a luminous body being only partly obscured by another body.
Perihelion: The point in an orbit closest to the body around which it orbits.
Period/periodic time (T): The time taken for an orbiting body to complete one circuit relative to the stars.
Plane: A flat surface.
Plane cosine rule: In any plane triangle, if a, b and c are the side lengths and C is the angle opposite the side of length c, then $c^2 = a^2 + b^2 - 2ab\cos C$.
Pole star: The well-known star close to the north celestial pole.
Precession: A slow movement, relative to the stars, of the equinoxes around the ecliptic.
Product: The result of multiplying two or more numbers.
Proportion: This occurs when two ratios are or can be equated. Also called the rule of three.
Pythagoras theorem: In a right-angled triangle the sum of the squares on the two shorter sides is equal to the square of the hypotenuse.

Quadrant boundaries: The angles 0°, 90°, 180°, 360°.
Quadruple: See Page 98.
Quintuple: Five quantities the sum of the squares of four of which is equal to the square of the fifth.
Quotient: The result of dividing on number by another.

Radian: The angle between two radii of a circle such that the length of the arc between them is equal to the radius of the circle is one radian.

Radius vector (r): The length of the line joining the Sun to a planet. Or the distance of any orbiting body from its primary.

Rational number: A number that is an integer of fraction.

Reciprocal: The result of dividing a number into one.

Rectangular coordinate system (x, y, z): A coordinate system in which the axes are mutually perpendicular.

Rectangular equatorial coordinates (x, y, z): The coordinates of a heavenly body in a coordinate system which uses three mutually perpendicular axes, where the line from the Sun or Earth to the First Point of Aries is the first of these, the second being in the equatorial plane and the third is defined by the direction of the north celestial pole

Right Ascension (R.A., α,): An angle measured from the First Point of Aries around the ecliptic to the meridian of a body. It is measured in hours and minutes (24 hours = 360°).

Scalene triangle: A triangle in which none of the sides are equal.

Semi-diameter: Half of the diameter.

Semimajor axis (a): In an elliptical orbit, half of the longer axis.

Semiminor axis (b): In an elliptical orbit, half of the shorter axis.

Series expansion: An infinite series representing a mathematical variable.

Sidereal day: The time taken by the Earth to make one rotation on its axis relative to the stars.

Sidereal year: The time taken by the Earth to make one revolution around the Sun relative to the stars.

Similar triangles: Triangles which are the same shape (their sides are proportional).

Solstice: The points at which the Sun reaches its greatest declination, north or south.

South celestial pole: The point where the line from the Earth's centre to the south pole meets the celestial sphere.

Spherical triangle: See Page 70.

Transcendental equation: An equation whose solution(s) are not the solution(s) of an algebraic equation. An algebraic equation is of the form $a_0x^n + a_1x^{n-1} + \ldots + a_n = 0$ where $a_0, a_1, \ldots, a_n$ are rational coefficients.

Triple: See Page 39.

True anomaly (v): See Pages 32, 55.

Umbra: A region of complete shadow

Vernal equinox: See First Point of Aries

Vinculum: A bar symbol put on top of a digit or group of digits to indicate that they are negative.

Zenith: The point on the celestial sphere which is directly overhead.

VEDIC SUTRAS

एकाधिकेन पूर्वेण
By One More than the One Before

निखिलं नवतश्चरमं दशतः
All from 9 and the Last from 10

उर्ध्वतिर्यग्भ्यामं
Vertically and Crosswise

परावर्त्य योजयेत्
Transpose and Apply

शून्यं साम्यसमुच्चये
If the Samuccaya is the Same it is Zero

आनुरूप्ये शून्यं अन्यत्
If One is in Ratio the Other is Zero

संकलन व्यवकलनाभ्यां
By Addition and by Subtraction

पूरणापूरणाभ्यां
By the Completion or Non-Completion

चलनकलनाभ्याम्
Differential Calculus

यावदूनं
By the Deficiency

व्यष्टिसमष्टिः
Specific and General

शेषाण्यङ्केन चरमेण
The Remainders by the Last Digit

सोपान्त्यद्वयमन्त्यं
The Ultimate and Twice the Penultimate

एकन्यूनेन पूर्वेण
By One Less than the One Before

गुणितसमुच्चयः
The Product of the Sum

गुणकसमुच्चयः
All the Multipliers

आनुरूप्येण
Proportionately

शिष्यते शेषसंज्ञः
The Remainder Remains Constant

आद्यमाद्येनान्त्यमन्त्येन
The First by the First and the Last by the Last

केवलैः सप्तकं गुण्यात्
For 7 the Multiplicand is 143

वेष्टनम्
By Osculation

यावदूनं तावदूनं
Lessen by the Deficiency

यावदूनं तावदूनीकृत्य वर्गं च योजयेत्
*Whatever the Deficiency lessen by that amount
and set up the Square of the Deficiency*

अन्त्ययोर्दशकेऽपि
Last Totalling 10

अन्त्ययोरेव
Only the Last Terms

समुच्चयगुणितः
The Sum of the Products

लोपनस्थापनाभ्यां
By Alternate Elimination and Retention

विलोकनं
By Mere Observation

गुणितसमच्चयः समुच्चयगुणितः
The Product of the Sum is the Sum of the Products

ध्वजाङ्क
On the Flag

INDEX